Thermodynamics - Kinetics of Dynamic Systems

Thermodynamics - Kinetics of Dynamic Systems

Editor

Federico Noguerra

scitus
academics

Thermodynamics - Kinetics of Dynamic Systems
Edited by **Federico Noguerra**

ISBN: 978-1-68117-467-9
Library of Congress Control Number: 2016942829

© 2017 by
SCITUS Academics LLC,
www.scitusacademics.com
Box No. 4766, 616 Corporate Way,
Suite 2, Valley Cottage,
NY 10989

Printed in United States of America on Acid Free Paper

Preface

Thermodynamics is the study of heat, "thermo," and work, "dynamics." Thermodynamics is the physics department that deals with the relationships between other forms of energy and heat. It describes in particular how thermal energy is converted into and from certain other types of energy and how it affects matter. It was founded in the 19th century when scientists first discovered how steam engines can be built and run. Thermodynamics only deals with the massive response of a method that we can observe and identify in experiments. The kinetic theory of gases discusses small-scale gas interactions. The methods complement one another; some principles are better understood in thermodynamics and some principles are better explained by the theory of kinetics. Thermodynamics is therefore concerned with a few properties of matter; heat is the most important of these. Heat is energy transferred between substances or systems due to a difference in temperature between them. Heat is preserved as a type of energy, i.e. cannot be created or possibly destroyed. It can, however, be transferred from one location to the next. Heat can also be converted to and from other energies.

Focusing on a wide range of thermodynamic applications, collects a series of contributions from the world's scientists, systematically gathered, this book will be of valuable for students and researchers who are interested in thermodynamics.

Table of Contents

CHAPTER 1

Kinetics and Thermodynamics of Protein Folding

Hongxing Lei[1, 2] and Yong Duan[2, 3]

[1] Beijing Institute of Genomics, Chinese Academy of Sciences, Beijing, China

[2] UC Davis Genome Center and Department of Applied Science,One Shields Avenue, Davis, USA

[3] College of Physics, Huazhong University of Science and Technology, Wuhan, China

1. INTRODUCTION

Proteins are the major functional elements in the living cells. Genetic information is stored in DNA. To release this information, DNA needs to be transcribed into mRNA which in turn is translated into protein. The alphabets are four bases for DNA and twenty amino acids for protein. The genetic code was revealed in 1950s with every amino acid coded by three consecutive bases. However, the amino acid sequence is only the primary structure of proteins. For proteins to be functional, the primary structure needs to fold into tertiary structure which is the optimal packing of secondary structures, namely alpha-helix and beta-sheet. In some cases, the tertiary structures of several proteins or subunits need to come together and form quaternary structure. The so-called "protein folding" problem mainly concerns the detailed physical transition process from primary structure to tertiary structure.

Protein folding mechanism consists of two major issues: kinetics and thermodynamics. Thermodynamically, the native state is the dominant and most stable state for proteins. Kinetically, however, nascent proteins take very different routes to reach the native state. Both issues have been extensively investigated by experimental as well as theoretical studies. The pioneering work by Christian Anfinsen in 1957 led to the creation and dominance of the "thermodynamic hypothesis" (also called "Anfinsen's dogma") which states that the native state is unique, stable and kinetically accessible free energy minimum (Anfinsen 1973). Under this guidance, many works have been done to pursue the illusive kinetically accessible folding pathways. One of the most famous earlier example is the "Levinthal's paradox" presented by Cyrus Levinthal in 1968, which states that the conformational space of proteins is so large that it will take forever for proteins to sample all the possible conformations before finding the global minimum (Levintha.C 1968). This essentially eliminated the possibility of global conformational search and pointed to the optimized folding pathways.

Towards this end, several well-known theories have been presented. The "framework theory" or similar "diffusion-collision theory" states that the formation of secondary structures is the first step and foundation of the global folding (Karplus and Weaver 1994). The "nucleation condensation theory", on the other hand, emphasizes the contribution from specific global contact as the initiation point of both secondary structure formation and global folding (Fersht 1995). In contrast to the emphasis on native contacts -- global or local— in these two theories, the "hydrophobic hydration theory" states that the general repulsion between hydrophobic residues and water environment drives the spatial redistribution of polar and non-polar residues and the eventual global folding (Dill 1990). In the more recent "funnel theory", the kinetics and thermodynamics of protein folding are better illustrated as funnel-shaped where both conformational space (entropy) and energy (enthalpy) gradually decrease and numerous kinetic traps exist en route to the global folding (Bryngelson, Onuchic et al. 1995). However, the driving force for protein folding is not specified in this theory.

In order to prove or disprove any theory, experimental evidence is needed. There are several techniques developed or applied to protein folding problem. First, the structure of the investigated protein needs to be solved by X-ray crystallography or NMR (nuclear magnetic resonance). High resolution X-ray structure is preferred. However, many of the model proteins can not be crystallized, therefore only NMR structures are available. Circular dichroism (CD) is one of the classical techniques for protein folding study. The proportion of secondary structures can be reflected in CD spectrum. The change of CD spectrum under different temperature or denaturant concentration can be used to deduce melting temperature or unfolding free energy. To study the fast folding process, however, CD itself is insufficient. Fast, time-resolved techniques include ultrafast mixing, laser temperature jump and many others. Other than CD, natural or designed fluorescence probes can be used to monitor the folding process. It should be noted that the "real" folding process may not be reflected with high fidelity in these folding experiments due to the artificial folding environment. In the living cells, proteins are synthesized on ribosome one residue at a time and the final products exist in a crowded physiological condition. In the folding experiments, however, proteins stay in free artificial solution and undergo various perturbations such as denaturation. In addition, fluorescence signal of specific probes can not be simply interpreted as the global protein folding, rather it only reflects the distance between the two selected residues.

Apart from experiments, computer simulation is another approach to study protein folding mechanism. In the early days, due to the limited computing resources, protein folding simulations were performed with extremely simplified models such as lattice models and off-lattice models where each residue is represented as a bead and the movement of the residues is restricted. This gradually evolved into models with more and more realistic main chain and side chain representations including the popular all-atom models in the present era. The treatment of the solvent environment has also been evolving. The solvent was ignored in the earlier simulations (*in vacuo* simulations). Continuum models with different levels of sophistication have been developed over the years, including the most simple linear distance dependent model and the modern generalized Born (GB) models and Poisson Boltzmann (PB) models (Onufriev, Bashford et al. 2004). With the continuous growing of computing power, explicit representation of water atoms has also been used in many folding simulations.

The first ever microsecond folding simulation was performed with explicit solvent in 1998 (Duan and Kollman 1998). With the help from a super computer cluster, this simulation was 100-1000 times longer than any other folding simulation at that time, thus stimulated great interest from the general public. In this simulation, the folding pathway to an intermediate state was observed. The folding rate of 4.2 μs predicted based on the simulation was highly consistent with later experimental finding of 4.3 μs. The success of this milestone work was followed by the highly publicized folding@home project and IBM blue gene project among many other works (Zagrovic, Snow et al. 2002). It should be noted that thousands or more simulation trajectories are utilized in the folding@home project which so far can not reach the time scale for protein folding. The limitation is due to the use of idle personal desktop computers which has far less computing power than super computers. The IBM blue gene project can overcome this problem by building extremely powerful computers that can cover millisecond folding simulation. However, it has yet to make a significant progress in protein folding. In a recently published work, a specially designed super computer succeeded in the folding of two small proteins (Shaw, Maragakis et al.). Although computing power does not seem to be the greatest hurdle from now on, this success is unlikely to extend broadly in the near future. The greater challenge lies in the accuracy of simulation force fields which will be discussed later.

In this chapter, we focus on theoretical studies of protein folding by molecular dynamics simulations. The kinetics of protein folding can be studies by conventional molecular dynamics (CMD). But the insufficient sampling in current CMD simulations prevents the extraction of thermodynamic information. This has prompted the development of enhanced sampling techniques, among which the most widely adopted technique is replica exchange molecular dynamics (REMD), otherwise called parallel tempering. In the past few years, we have applied both CMD and REMD to the *ab initio*folding – meaning folding from extended polypeptide chain without any biased force towards the native contacts – of several model proteins, including villin headpiece subdomain (HP35), B domain of protein A (BdpA), albumin binding domain (ABD), and a full sequence design protein (FSD). To enhance the conformational sampling, we used an implicit solvent model GB/SA (surface area) implemented in the AMBER simulation package. The accuracy of protein folding reached sub-angstrom in most of these simulations, a significant improvement over previous simulations. Based on these high accuracy simulations, we were able to investigate the kinetics and thermodynamics of protein folding. The summary of our findings will be presented here in details. Finally, we will stress the critical role of force filed development in studying folding mechanism by simulation.

2. KINETICS AND THERMODYNAMICS FROM *AB INITIO* FOLDING SIMULATIONS

2.1. Villin Headpiece Subdomain: Traditional Analysis

Villin headpiece subdomain (HP35) is a small helical protein (35 residues) with a unique three helix architecture (Fig 1). Helix I is nearly perpendicular to the plane formed by helices II and III. The three-dimensional structure was solved earlier by an NMR experiment and more recently by a high resolution X-ray experiment (Chiu, Kubelka et al. 2005). Due to the small size and rich structural information, HP35 has attracted a lot of attention from both experimentalists and theoreticians.

In our previous works, we have conducted CMD to study the folding pathway and REMD to study the thermodynamics of HP35 (Lei and Duan 2007; Lei, Wu et al. 2007). In the CMD work, we observed two intermediate states from the twenty folding trajectories (1μs each), one with the well-folded helix II/III segment (defined as the major intermediate state) and the other with the well-folded helix I/II segment (defined as the minor intermediate state). The best folded structure had C_α-RMSD of 0.39 Å and the most representative folded structure had C_α-RMSD of 1.63 Å. The productive folding always went through the major intermediate state while no productive folding was observed through the minor intermediate state. Further examination revealed that the initiation of the folding was around the second turn between Phe17 and Pro21 rather than the hydrophobic core formed by Phe6/Phe10/Phe17. On the other hand, Gly11 was likely most accountable for the flexibility of helix I. In addition, the high occupancy of short-distance native contacts and low occupancy of long-distance native contacts pointed to the importance of local native contacts to the fast folding kinetics of HP35.

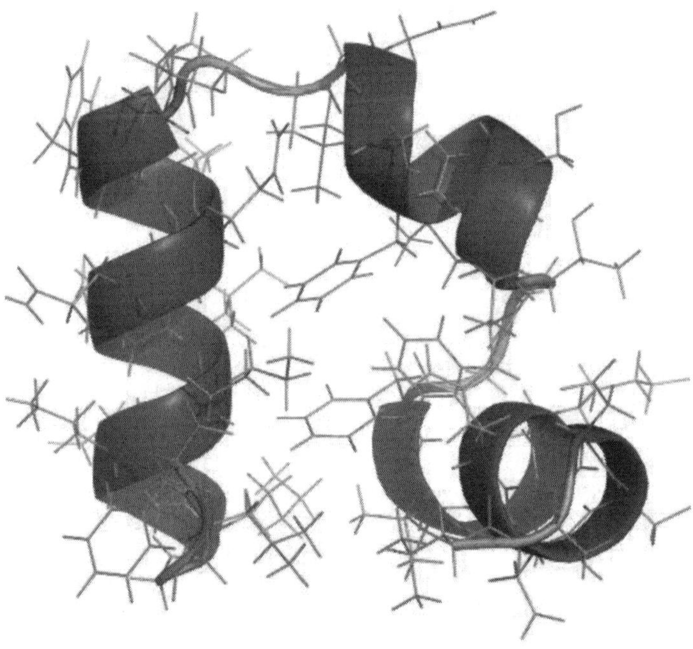

FIGURE 1. Structure of villin headpiece subdomain (HP35)

In the REMD work, we conducted two sets of REMD simulations (20 replicas and 200 ns for each replica) with convergent results. The best folded structure had C_α-RMSD of 0.46 Å and the most representative folded structure had C_α-RMSD of 1.78 Å. The folding landscape of HP35 was partitioned into four thermodynamic states, namely the denatured state, native state, and the two aforementioned intermediate states. The dynamic feature of the folding landscape at selected temperatures (300 K, 340 K and 360 K) was consistent in both REMD simulations and the corresponding CMD simulations. A major free energy barrier (2.8 kcal/mol) existed between the denatured state and the major intermediate state, while a minor free energy barrier (1.3 kcal/mol) existed between the major intermediate state and the native state. In addition, a melting temperature of 339 K was predicted from the heat capacity profile, very close to the experimentally determined melting temperature of 342 K.

Because of the small size, HP35 has been considered as a classical two-state folder. This notion is supported by some earlier folding experiments. However, our simulation clearly pointed to the existence of folding intermediates. Our two-stage folding model is supported by some more recent folding experiments. In a laser temperature-jump kinetic experiment, the unfolding kinetics was fit by a bi-exponential function, with slow (5 μs) and fast (70 ns) phases. The slower phase corresponds to the overall folding/unfolding, and the fast phase was due to rapid equilibration between the native and nearby states. In a solid-state NMR study, three residues (Val9, Ala16, and Leu28) from the three helices exhibited distinct behavior during the denaturation process, and a two-step folding mechanism was proposed. In an unfolding study using fluorescence resonance energy transfer, Glasscock and co-workers demonstrated that the turn linking helices II and III remains compact under the denaturation condition (Glasscock, Zhu et al. 2008), suggesting that the unfolding of HP35 consists of multiple steps and starts with the unfolding of helix I. In a mutagenesis experiment, Bunagan et al. showed that the second turn region plays an important role in the folding rate of HP35 (Bunagan, Gao et al. 2009). A recent freeze-quenching experiment by Hu and co-workers revealed an intermediate state with native secondary structures and nonnative tertiary contacts (Hu, Havlin et al. 2009). These experiments are highly consistent with our observations in terms of both the stepwise folding and the rate-limiting step. Kubelka et al. proposed a three-state model in which the interconversion between the intermediate state and folded state is much faster than that between the intermediate state and the unfolded state (Kubelka, Henry et al. 2008). Therefore, the intermediate state lies on the folded side of the major free energy barrier, which is consistent with the separation of the unfolded state from the other states in our folding simulation. The estimation of 1.6–2.0 kcal/mol for the major free-energy barrier is also consistent with the estimation from our previous REMD simulation.

Nevertheless, controversy still exists regarding the folding mechanism of this small protein. In a recent work by Reiner et al., a folded segment with helices I/II was proposed as the intermediate state (Reiner, Henklein et al.), which corresponds to the off-pathway minor intermediate state in our work. It should be noted that different perturbations to the system, including high concentrations of denaturant, high temperatures, and site mutagenesis, have been utilized in different folding experiments. Because of the small size of HP35, the folding process may be sensitive to some of these perturbations. With the continuous development of experimental techniques that allow minimal perturbation and monitoring of the folding process at higher spatial and temporal resolution, the protein-folding mechanism will become more and more clear.

2.2. Villin Headpiece Subdomain: Network Analysis

REMD is one of the most efficient sampling techniques for protein folding. However, due to the non-physical transitions from the exchange of conformations at different temperatures, its usage is mostly restricted to thermodynamics study. To get better understanding of the kinetics, we decided to extend the CMD simulations from the previous 1 μs to 10 μs in five selected simulation trajectories (Lei, Su et al.). Consistent with REMD, the folding free energy landscape displayed four folding states (Fig 2), the denatured state on the upper right region, the native state on the lower left region, the major intermediate state on the lower right region, and the minor intermediate state on the upper left region. The construction of the 2D landscape was based on two selected reaction coordinates, RMSD of segment A (helix I/II) and segment B (helix II/III). All five trajectories were combined together, and the population of each conformation in a small zone was converted to free energy by log transformation. From the folding landscape, we can see focused sampling in the native state, sparse sampling in the minor intermediate states, and heterogeneous sampling in the denatured state and the major intermediates state. The heavy sampling in the denatured state was likely due to the limited simulation trajectories. Ideally, thousands of trajectories are needed to reach good sampling. However, long simulations like this one are computer intensive beyond the capacity of a typical institution.

The above-described 2D landscape is only an overall display of the conformational sampling. To get more details, we performed conformational clustering based on the combined five trajectories. We here use the top ten most populated conformational clusters to describe the conformational sampling (Fig 3). The center of each conformational cluster was used to represent the cluster. Among the top ten clusters, we can see three conformations in the native state (clusters 2, 3 and 10, colored in purple), three conformations in the major intermediate state (clusters 5, 6 and 9, colored in green), and four conformations in the denatured state (clusters 1, 4, 7 and 8, colored in blue), while the minor intermediate state did not show up due to small overall population. The overall energy (enthalpy) was not a good indication of the folding. In fact, the energy of the native state conformations was the highest and that of the denatured state conformations was the lowest. This observation did not violate the "thermodynamics hypothesis" because the conformational entropy was not included in the energy calculation. Entropy evaluation has long been a difficult subject in the field of computational biochemistry. A breakthrough will extend the application of force fields to protein structure prediction.

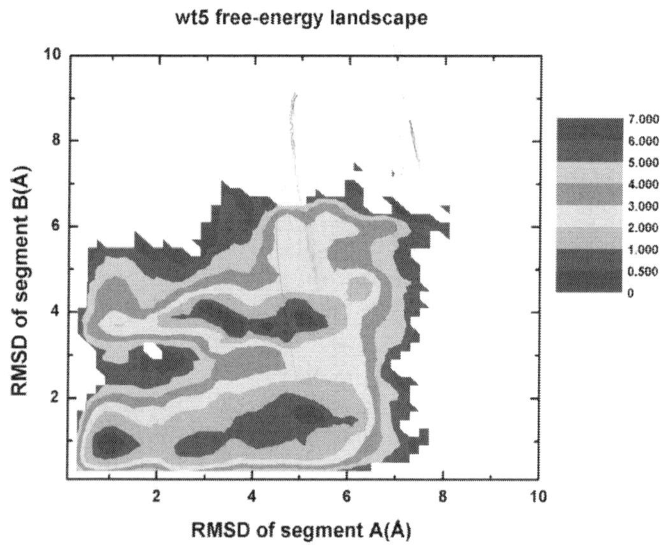

FIGURE 2. Folding free energy landscape of HP35

Based on conformational clustering, we can study the kinetics and thermodynamics of protein folding using a new technique called network analysis. Traditionally, protein folding is illustrated by 1D profiles such as RMSD (global or partial), energy, solvent accessible surface area, radius of gyration and selected distances. The hyper-dimensional nature of protein folding makes none of these 1D profiles adequate to reflect the folding process. The emergence of 2D maps such as the one in Fig 2greatly alleviate the problem by combining two independent profiles in one map. However, 2D maps are still insufficient to represent the hyper-dimensional process. Under this circumstance, several novel approaches have been applied to protein folding in recent years, including the disconnectivity graph by Karplus and network analysis pioneered by Caflisch (Krivov and Karplus 2004; Caflisch 2006).

Network analysis has gained popularity in protein folding recently (Bowman, Huang et al.; Jiang, Chen et al.). In network analysis, protein conformations are represented as nodes and the transitions among different conformations are represented as edges. Both nodes and edges can be colored based on a specified property, and analysis can be done based on the topological distribution of conformations with a specified property. In the folding network of the combined five trajectories (Fig 4), we painted the nodes according to the state identity of the conformation and displayed the structure of the top ten populated conformations. From this network, we can see the clear separation of the denatured state from the native state and major intermediate state. The minor intermediate state was also connected to the denatured state. These findings were consistent with the observation from the 2D maps. A new finding is the mixing of the native state and major intermediate state which were clearly separated in the 2D map. The implication of this new finding is that the barrier between these two states is so small that they can easily convert to each other, which is supported by experimental evidence. This study demonstrated the power of network analysis and suggested more caution on interpreting 2D maps of protein folding.

The global folding network better reflect the thermodynamics of protein folding. To understand the kinetics of protein folding better, a simplified network with shortest path can be constructed (Fig 5). In this network, the shortest path connecting the denatured state, the major intermediate state and the native state was extracted from the global network. A clear flow of conformational transition from the denatured state to the major intermediate state and then to the native state was demonstrated in this network. Even the number of transitions between any neighboring conformations can be labeled on the network. In the denatured state, there were three short paths from the four top conformations to the major intermediate state, suggesting multiple folding pathways. Two conformations in the minor intermediate state were embedded in the denatured state, suggesting them as off-pathway intermediate. In the major intermediate state, the two top conformations close to the denatured state (clusters 6 and 9 in Fig 3) had wrongly folded segment A, while the top conformation close to the native state (cluster 5 in Fig 3) had a near native structure. This information on the intra-state conformational transition is also helpful to reveal the details in the protein folding process. In the native state, the high connectivity among the conformations within the state and also with the major intermediate state suggests the relative independence of the native state and the low barrier between the native state and the major intermediate state.

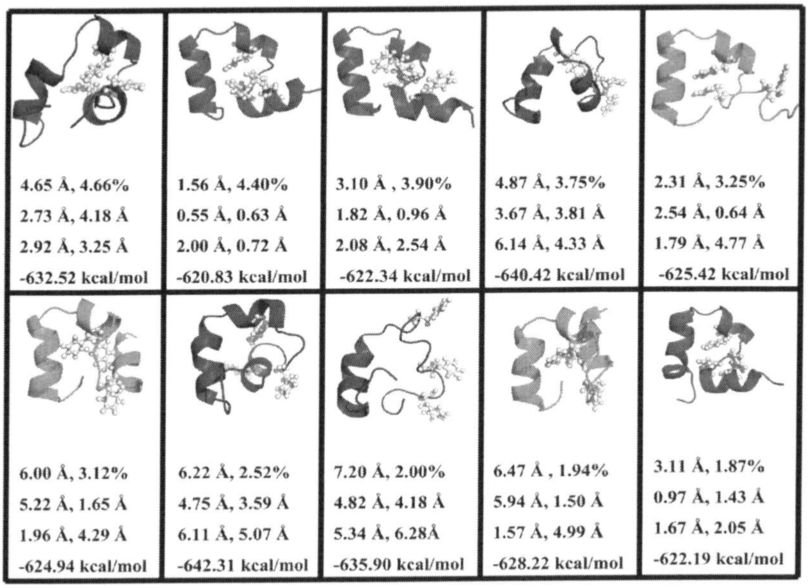

FIGURE 3. Representative structures of the top ten populated clusters of HP35

In the above two sub-sections, we have presented our study of folding mechanism for HP35 wild type. A challenging problem in this field is whether mutational effect can be reproduced in simulation. To enhance the folding rate of HP35 wild type, a mutant was designed to replace two partially buried lysine residues with non-natural neutral residues which resulted in the sub-microsecond folding. We conducted similar simulations for this HP35 mutant and compared with that of the wild type. Similar to the wild type, the mutant simulation also reached sub-angstrom accuracy (Lei, Deng et al. 2008). The folding free energy landscape also displayed similar feature with four folding states. However, some difference was also observed, especially the increased population of the native state, the decreased population of the denatured state, higher melting temperature, and the lower free energy barrier between the denatured state and the major intermediate state. These pointed to higher stability of the native state and faster folding which is consistent with the experiment. A surprising finding is the folding pathways through both intermediate states. Therefore, the two mutated residues not only stabilized the local secondary structure (helix III), but also reshaped the folding landscape in several different ways. This kind of detailed information can not be obtained from folding experiment as yet. Thus, computer simulation will play a complimentary role in the understanding of folding mechanism in the foreseeable future.

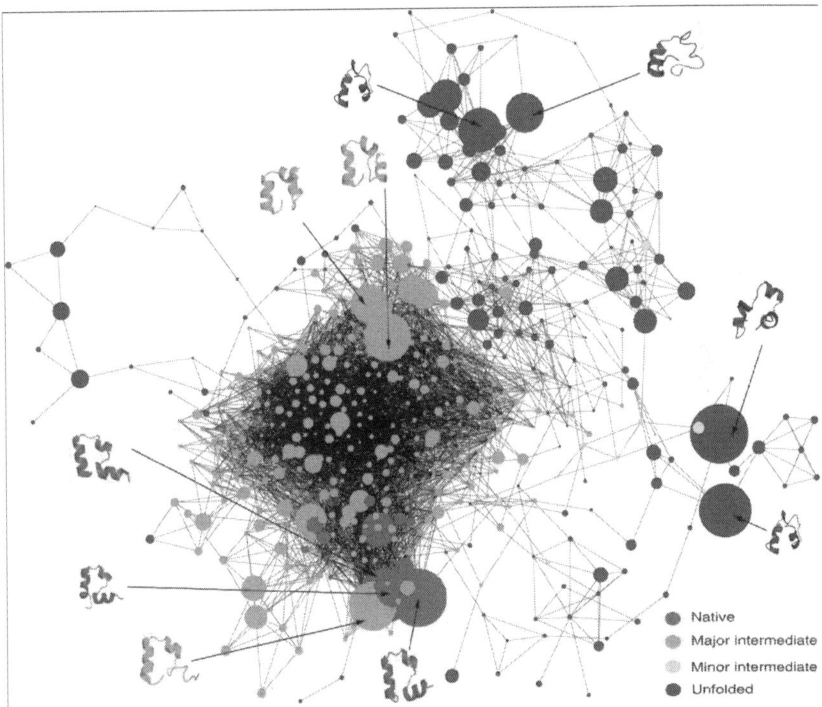

FIGURE 4. Folding network of HP35

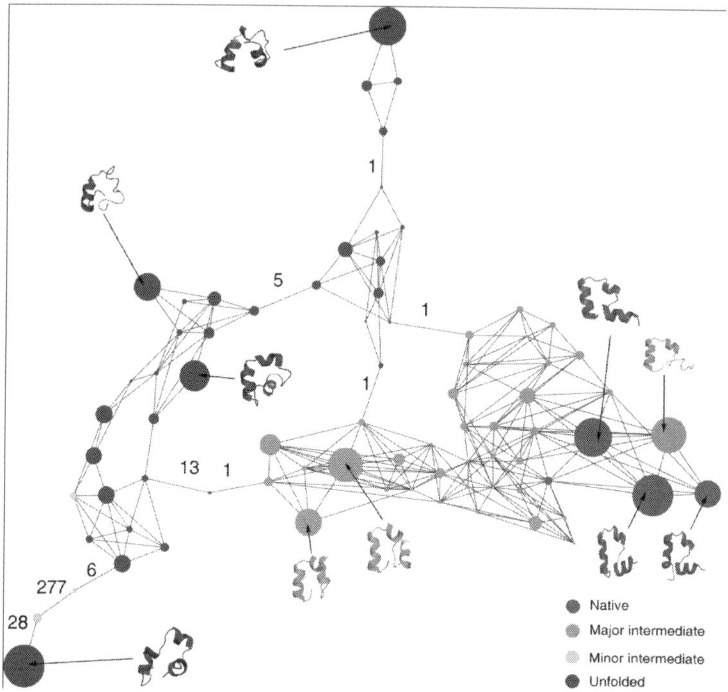

FIGURE 5. A simplified folding network of HP35

2.3. Folding of Three Other Model Proteins

In addition to the folding studies of HP35 wild type and mutant, we also conducted ab initio folding on three other model proteins, namely B domain of protein A (BdpA), albumin binding domain (ABD), and a full sequence design protein (FSD). Here we will briefly describe our results. For detailed information, please refer to the original publications. BdpA is another three-helix protein with a different tertiary architecture (Fig 6). Helices II and III are relatively parallel to each other and form a plane. Helix I docks to this plane with a tilt angle relative to the other two helices. BdpA has 60 residues in the full length version and 47 residues in the truncated version (residues 10-56) where the unstructured terminal residues are trimmed off. In our simulation work, ab initio folding on both versions has been conducted.

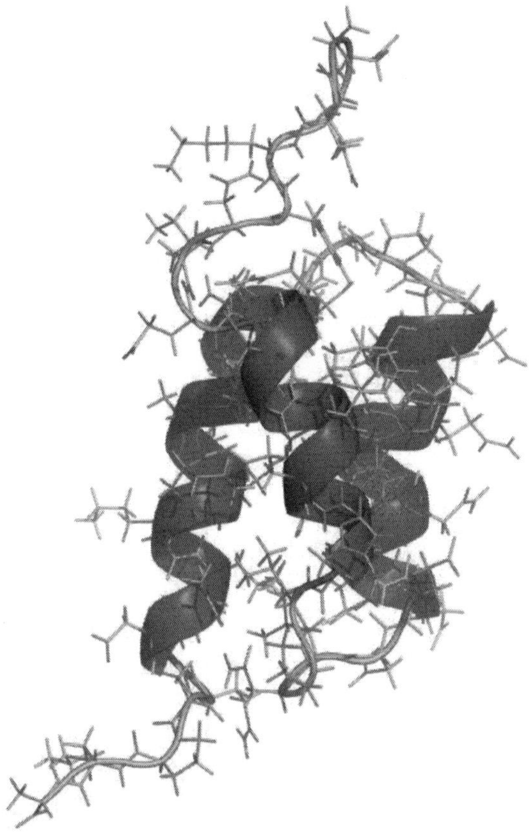

FIGURE 6. Structure of B domain of protein A (BdpA)

Successful folding was achieved in our simulation (Lei, Wu et al. 2008). The best folded structure was 0.8 Å RMSD in the CMD of truncated version and 1.3 Å in the REMD of full length version. In the CMD simulations, the folding initiated from the formation of helix III, followed by the folding of a intermediate state with well folded helix II/III segment, and completed with the docking of helix I to the helix II/III segment. The folding pathway was similar to that of HP35 except for the initiation step, where it was the formation of helix III for BdpA and formation of the second turn for HP35. In the REMD simulations, the most populated conformation was a folded conformation with 64.1% population. The melting temperature of 362 K from the heat capacity profile was also close to the experimentally derived melting temperature of 346 K. From the calculated potential of mean force, the native state was 0.8 kcal/mol favored over the denatured state, and the free energy barrier from the denatured state to the native state was 3.7 kcal/mol. These findings were in qualitative agreement with folding experiments of BdpA. In addition, we tested the structure prediction performance based on AMBER potential energy and DFIRE statistical energy. Both gave similar performance with most predicted structures near 3.0-3.5 Å RMSD, while the structure with the lowest AMBER potential energy was only 2.0 Å RMSD.

Albumin binding domain (ABD) is yet another three helix protein with different topological feature from both HP35 and BdpA (Fig 7). Helices I and III are relatively long and parallel with each other, while the shorter helix II serves as linker. NMR studies revealed high uncertainty in the first loop and dynamics around helix II. The experimentally determined folding rate for wild type ABD is 6 μs. Enhanced folding rate was achieved in two mutants (2.5 μs for K5I and 1.0 μs for K5I/K39V).

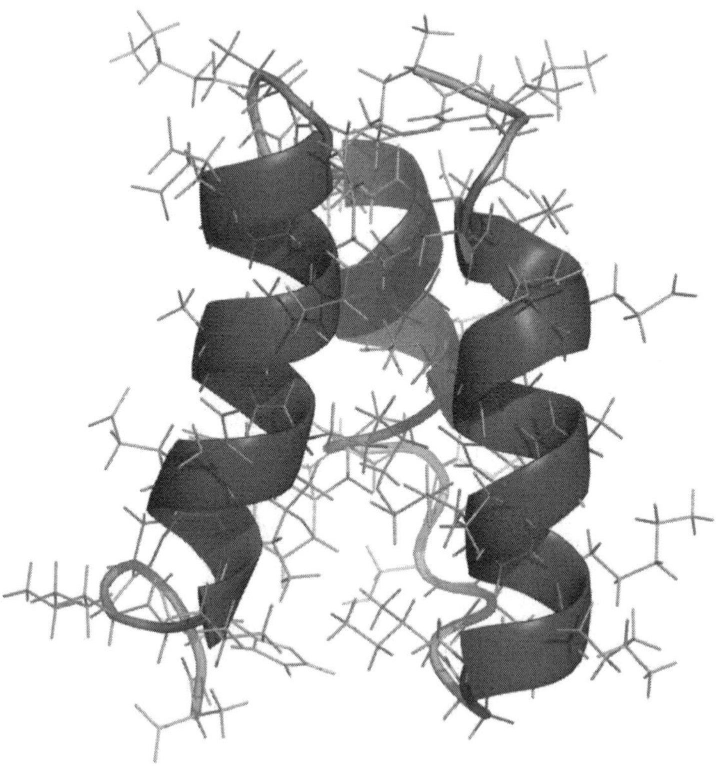

FIGURE 7. Structure of albumin binding domain (ABD)

We conducted 20 CMD simulations (400 ns each) for each of the above mentioned three ABD variants. Although the size of ABD is comparable to the truncated BdpA (both 47 residues), no successful folding of ABD was reported prior to our study. In our simulations, the best folded structure reached 2.0 Å RMSD, indicating the first ever successful folding of ABD (Lei and Duan 2007). The folding started from the formation of helix I, followed by the formation of the other two helices, and completed by the optimal packing of the three helices. Although two hydrophobic cores exist in the middle, their formation was late in the simulation, suggesting that they are not the driving force of the global folding. Examination of conformational sampling revealed that the folded conformation was the most populated and significant formation of helices also appeared in other populated conformations. Compared to HP35 and BdpA, the accuracy of folding was lower for ABD. This was likely coming from several sources. First, ABD is highly dynamic according to NMR experiments which makes it difficult to choose a reference structure to determine folding accuracy. Second, the trajectory length for ABD (400 ns) was significantly shorter than that of HP35 and BdpA (both 1 μs). Third and likely most importantly, some features in ABD can not be modeled well in the current simulation force fields. The hydrophobic core in ABD may play important role in the folding mechanism, while it may play minor role in the folding of HP35 and BdpA. Therefore, inaccurate modeling of hydrophobic interaction will lead to less accurate folding of ABD than that of HP35 and BdpA. In addition, helix boundaries were slightly shifted in the simulation compared with the NMR structure, which may be partially due to the inaccurate parameterization of certain amino acids such as Valine.

The three model proteins described above are all helical proteins. A more challenging task is to fold proteins with both alpha-helix and beta-sheet secondary structures. Full sequence design protein (FSD) is a designed 28-residue α/β protein. Our previous attempts on the ab initio folding of FSD were unsuccessful with the same simulation force field (AMBER FF03) used in the successful folding of the three helical proteins. Therefore, we decided to re-parameterize the force field under the same solvation scheme (GB/SA) for a better balance of the two major secondary structures.

Using the newly developed force field, we conducted ab initio folding of FSD with both CMD and REMD (Lei, Wang et al. 2009). High accuracy folding was achieved in terms of both the best folded structure (0.8 Å RMSD) and the population of the folded conformation (64.2%). High diversity was observed in the sampled denatured conformations, including a long helix, a helix hairpin and a long beta-hairpin, indicating good balance of the two major secondary structures. The folding of FSD followed two distinctive pathways. The major pathway began with the formation of the helix, while the minor pathway started with the formation of the beta-hairpin. More specifically, the initiation of the helix started from the C-terminal and propagated to the N-terminal. The free energy profiles showed different stability for the two structural elements. For the helix segment, the native helical structure had significantly lower free energy than other conformations. For the hairpin segment, however, 2-3 non-native conformations existed with similar free energy, which led to several local traps on the free energy landscape. Kinetically, the free energy barrier was similar for the folding of both segments (2-3 kcal/mol), but it was a single barrier for the helix and multiple barriers for the hairpin. The melting temperature extracted from the heat capacity profile was 360 K. However, this temperature merely reflected the melting of the helix (~50% helicity at 360 K), while the population of globally folded conformation was close to 0% at 360 K. Therefore, caution should be taken when interpreting the "melting temperature" extracted from the heat capacity profile.

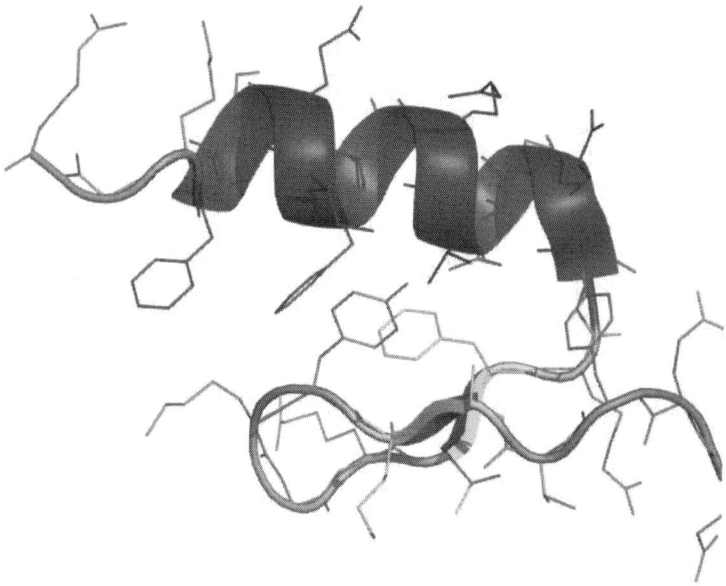

FIGURE 8. Structure of a full sequence design protein (FSD)

3. KINETICS AND THERMODYNAMICS FROM UNFOLDING SIMULATIONS

Ideally, protein folding mechanism should be studied by ab initio folding. However, due to the limited access of super computers by most research groups, ab initio folding simulations are limited to very few small model proteins such as listed in Section 2. For most other proteins, an alternative approach is unfolding simulation where the native starting structures gradually unfold under the high temperature. A major assumption with this approach is that folding pathway is the reverse of unfolding pathway. To validate this approach, we conducted unfolding simulations on HP35, BdpA and FSD and compared the unfolding pathways with the folding pathways from ab initio simulations. Here we mainly use the unfolding of HP35 as an example.

We conducted ten unfolding simulations of HP35 at 350 K (100 ns each) and used the average properties from these simulations to describe the unfolding process. One of the main findings from the ab initio folding simulations was the major intermediate state with well-folded helix II/III segment. Thus, we first examined the unfolding of the two structural segments (Fig 9). We can clearly see the faster unfolding of the helix I/II segment which reached complete unfolding before 20 ns. On the other hand, the unfolding of helix II/III segment was much slower and was still fluctuating after unfolding. The slower unfolding and higher stability of helix II/III segment suggest that it folds earlier than helix I/II segment, which is consistent with the finding from ab initio folding.

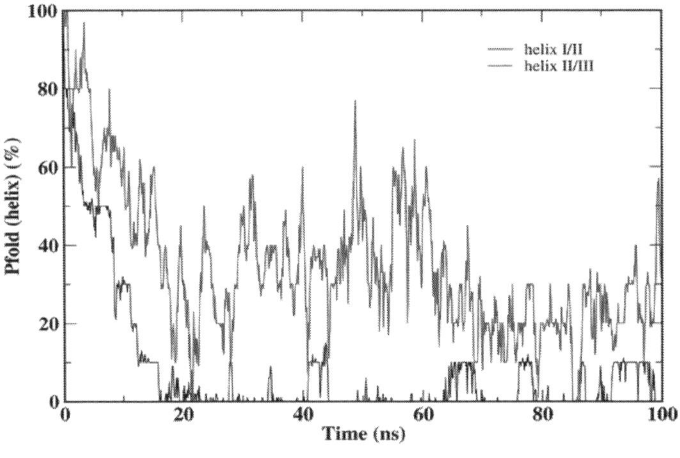

FIGURE 9. Unfolding of the two segments of HP35

Second, we examined the unfolding of the three individual helices using a simple helicity measurement (Fig 10). The three helices showed distinctive unfolding features. Helix I was the fastest to unfold and the least stable one, down to 50% within 10 ns and towards 25% near 100 ns. On the other hand, helix III was the slowest to unfold and the most stable one, fluctuating between 70% and 85% during the whole simulation time. This suggests that helix III is the first to fold and helix I is the last to fold, which is also consistent with the ab initio folding simulation.

Further evaluation of the unfolding can be performed at the residue level. We calculated the root mean square fluctuation (RMSF) for each residue during the entire unfolding process (Fig 11). Overall, the two terminal regions displayed highest fluctuation and the second half of the protein was significantly more stable than the first half. Heterogeneity was observed within the helices. Within helix I (residues 3-9), the middle residues 6 and 5 had the lowest fluctuation. Within helix II (residues 14-19), the C-terminal residues 17-19 had the lowest fluctuation. Within helix III (residues 22-31), most residues had the low fluctuation especially residues 24-30. Another interesting observation is the low fluctuation of residue 20 at the second turn. All these observations were consistent with the folding mechanism from the ab initio folding simulations.

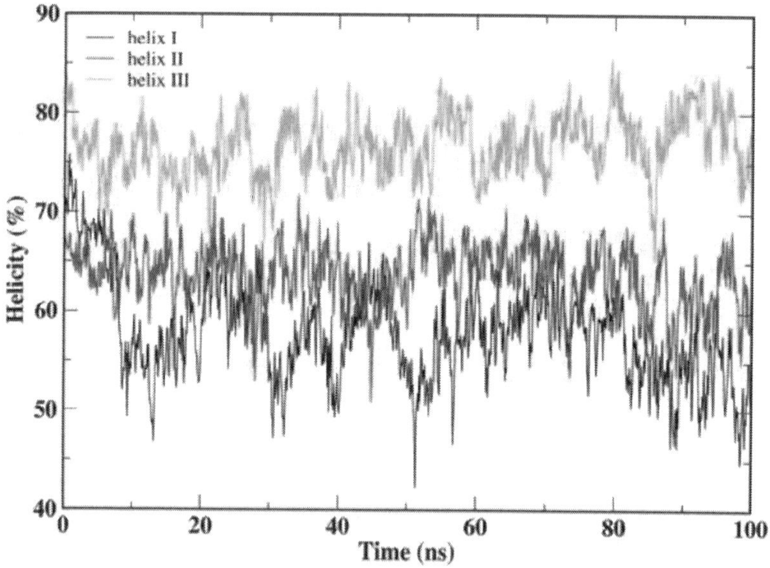

FIGURE 10. Unfolding of the three helices of HP35

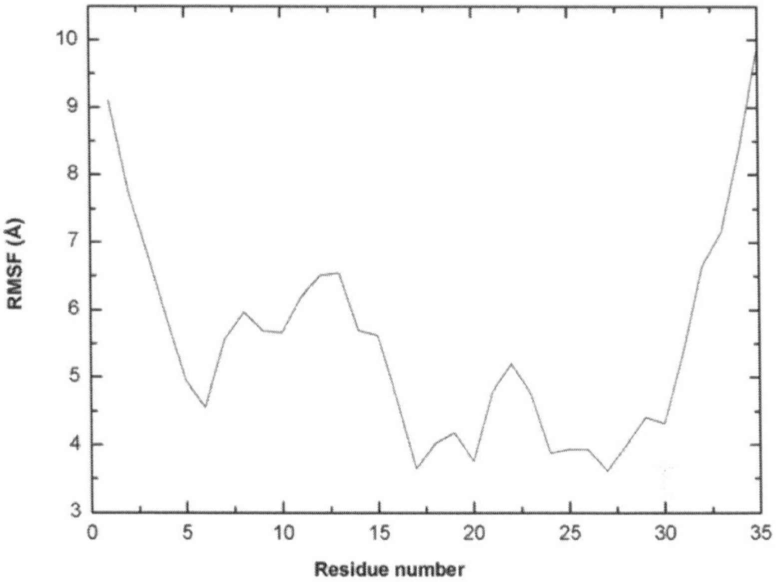

FIGURE 11. Dynamic feature of each residue in the unfolding of HP35

A more intuitive way to visualize the folding pathways is by constructing the folding landscape. We divided the 100 ns unfolding time into five time frames and constructed folding landscape during each time frame (the first and last time frames shown in Fig 12). During the first time frame (0-20 ns), the native state was the most dominant one, while unfolding to the major intermediate state and denatured state was also observed. During the last time frame (80-100 ns), the denatured state became the most dominant one, while the major intermediate state was observed but the native state and the minor intermediate state were almost undetectable. The reverse of this observed process was exactly what we observed in the ab initio folding of HP35.

In addition to the unfolding of HP35, we also conducted unfolding simulations on BdpA and FSD, all with ten trajectories of 100 ns simulations at 350 K. The unfolding mechanism from these two sets of simulations was also consistent with the folding mechanism from previously described ab initio folding simulations. In summary, our comparison between unfolding and ab initio folding suggests that unfolding is a valuable approach to study folding mechanism when ab initio folding is unfeasible. However, our comparison was based on three small model proteins, whether this conclusion can be extended to other proteins remains to be a question for future investigation.

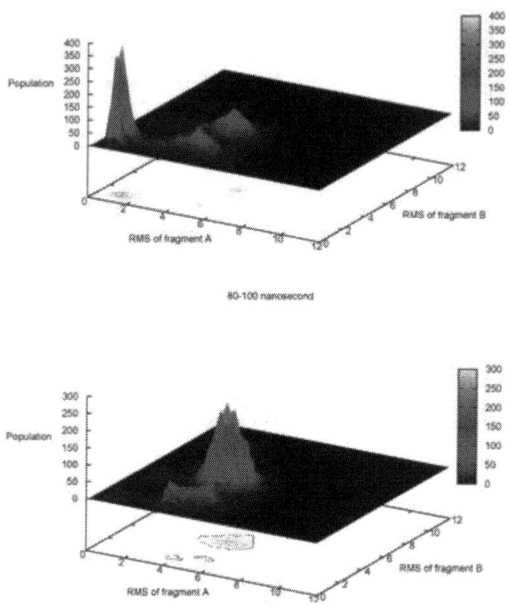

FIGURE 12. Shifting of structural ensembles during the unfolding of HP35

4. SIMULATION FORCE FIELD DEVELOPMENT

At this point, we should stress the importance of force field in protein folding simulation. Under appropriate protocol, a simulation is as good as the underlying force field is. Currently, the main stream force fields, namely AMBER, CHARMM, GROMOS and OPLS, are all point charge models (Duan, Wu et al. 2003). Under this philosophy, a partial charge is assigned to every atom of a specific amino acid with the overall charge reflecting the charge nature of the amino acid (+1, -1 or 0). In some schemes, the main chain atoms (N, H, CA, C and O) are restricted to have the same set of partial charges for every amino acid, while this restriction was not applied in some other schemes. The partial charges can be derived from fitting to the electrostatic potential calculated by quantum mechanics, or can be assigned by chemical intuition. Another major parameterization of force field is the torsion angles, both main chain and side chain, which is usually fitted to reflect the potential energy surface of amino acid analogs especially Alanine dipeptide calculated by quantum mechanics. The major advantage of the current generation force field is the speed. However, as longer and longer simulations being conducted, more and more problems have been revealed regarding the accuracy of these force fields, including significant bias towards a specific secondary structure (alpha-helix or beta-sheet). In light of these problems, the concept of polarizable force field has emerged. The major philosophical difference from point charge models is the dynamics in charge distribution. Polarizable force fields are being developed and will likely become the next generation force field soon.

5. CONCLUSION

Computer simulation is a powerful tool to study the kinetics and thermodynamics of protein folding. Here we summarized our study of folding mechanism on four model proteins by CMD an REMD. We have reached sub-angstrom folding on HP35, BdpA and FSD and 2.0 Å RMSD folding on ABD. From the high quality folding simulations, we extracted a plethora of information regarding the folding mechanism, including folding pathways, folding states, free energy barriers, melting temperature and folding landscape. We have also applied network analysis to the study of folding mechanism and revealed new information about the folding of HP35. In addition, the high consistency between unfolding simulations and *ab initio* folding simulations suggest that unfolding simulations can be used as an alternative.

NAN. ACKNOWLEDGEMENTS

This work was supported by research grants from NIH (Grants GM79383 and GM67168 to YD), NSFC (Grant 30870474 to HL) and SRF for ROCS, SEM (to HL). Usage of AMBER and Pymol, GRACE, VMD, Matlab and Rasmol graphics packages are gratefully acknowledged.

REFERENCES

1. C. B. Anfinsen, 1973 "Principles that govern the folding of protein chains." Science 181 96 22330 .

2. G. R. Bowman, X. Huang, et al. 2010 "Network models for molecular kinetics and their initial applications to human health." Cell Res 20 6 62230 .

3. J. D. Bryngelson, J. N. Onuchic, et al. 1995 "Funnels, Pathways, and the Energy Landscape of Protein-Folding- a Synthesis." Proteins-Structure Function and Genetics 21 3 167195 .

4. M. R. Bunagan, J. Gao, et al. 2009 "Probing the folding transition state structure of the villin headpiece subdomain via side chain and backbone mutagenesis." J Am Chem Soc 131 21 74706 .

5. A. Caflisch, 2006 "Network and graph analyses of folding free energy surfaces." Current Opinion in Structural Biology 16 1 7178 .

6. T. K. Chiu, J. Kubelka, et al. 2005 "High-resolution x-ray crystal structures of the villin headpiece subdomain, an ultrafast folding protein." Proc Natl Acad Sci U S A 102 21 7517-22 .

7. K. A. Dill, 1990 "Dominant forces in protein folding." Biochemistry 29 31 7133-55 .

8. Y. Duan, P. A. Kollman, 1998 "Pathways to a protein folding intermediate observed in a 1-microsecond simulation in aqueous solution." Science 282 5389 740-4 .

9. Y. Duan, C. Wu, et al. 2003 "A point-charge force field for molecular mechanics simulations of proteins based on condensed-phase quantum mechanical calculations." J Comp Chem 24 16 19992012 .

10. A. R. Fersht, 1995 "Optimization of rates of protein folding: the nucleation-condensation mechanism and its implications." Proc Natl Acad Sci U S A 92 24 10869-73 .

11. J. M. Glasscock, Y. Zhu, et al. 2008 "Using an amino acid fluorescence resonance energy transfer pair to probe protein unfolding: application to the villin headpiece subdomain and the LysM domain." Biochemistry 47 42 11070-6 .

12. K. N. Hu, R. H. Havlin, et al. 2009 "Quantitative determination of site-specific conformational distributions in an unfolded protein by solid-state nuclear magnetic resonance." J Mol Biol 392 4 1055-73 .

13. X. Jiang, C. Chen, et al. 2010 "Improvements of network approach for analysis of the folding free-energy surface of peptides and proteins." J Comput Chem 31 13 2502-9 .

14. M. Karplus, D. L. Weaver, 1994 "Protein folding dynamics: the diffusion-collision model and experimental data." Protein Sci 3 4 650-68

15. S. V. Krivov, M. Karplus, 2004 "Hidden complexity of free energy surfaces for peptide (protein) folding." Proceedings of the National Academy of Sciences of the United States of America 101 41 1476614770 .

16. J. Kubelka, E. R. Henry, et al. 2008 "Chemical, physical, and theoretical kinetics of an ultrafast folding protein." Proc Natl Acad Sci U S A 105 48 18655-62 .

17. H. Lei, X. Deng, et al. 2008 "The fast-folding HP35 double mutant has a substantially reduced primary folding free energy barrier." J Chem Phys 129 15 155-104 .

18. H. Lei, Y. Duan, 2007 "Ab initio folding of albumin binding domain from all-atom molecular dynamics simulation." J Phys Chem B 111 19 5458-63 .

19. H. Lei, Y. Duan, 2007 "Two-stage folding of HP-35 from Ab initio simulations." J Mol Biol 370 1 196206 .

20. H. Lei, Y. Su, et al. 2010 "Folding network of villin headpiece subdomain." Biophys J 99 10 337484 .

21. H. Lei, C. Wu, et al. 2007 "Folding free-energy landscape of villin headpiece subdomain from molecular dynamics simulations." Proc Natl Acad Sci U S A 104 12 492530 .

22. H. X. Lei, Z. X. Wang, et al. 2009 "Dual folding pathways of an alpha/beta protein from all-atom ab initio folding simulations." Journal of Chemical Physics 131 16):-.

23. H. X. Lei, C. Wu, et al. 2008 "Folding processes of the B domain of protein A to the native state observed in all-atom ab initio folding simulations." Journal of Chemical Physics 128 23):-.

24. Levintha.C 1968 "Are There Pathways for Protein Folding." Journal De Chimie Physique Et De Physico-Chimie Biologique 65 1 4445 .

25. A. Onufriev, D. Bashford, et al. 2004 "Exploring protein native states and large-scale conformational changes with a modified generalized born model." Proteins 55 2 383394 .

26. Reiner, P. Henklein, et al. 2010 "An unlocking/relocking barrier in conformational fluctuations of villin headpiece subdomain." Proc Natl Acad Sci U S A 107 11 495560 .

27. D. E. Shaw, P. Maragakis, et al. 2010 "Atomic-level characterization of the structural dynamics of proteins." Science 330 6002 3416 .

28. Zagrovic, C. D. Snow, et al. 2002 "Simulation of folding of a small alpha-helical protein in atomistic detail using worldwide-distributed computing." J Mol Biol 323 5 927-37 .

2

CHAPTER

Dynamics and Thermodynamics of Nanoclusters

Karo Michaelian [1,†,*] and Ivan Santamaría-Holek [2,†]

[1] *Instítuto de Física, Universidad Nacional Autónoma de México, Apdo. Post. 20-364, México D.F. 01000, Mexico*

[2] *UMDI-Facultad de Ciencias, Universidad Nacional Autonoma de Mexico, Campus Juriquilla, Queretaro 76230, Mexico*

ABSTRACT

The dynamic and thermodynamic properties of nanoclusters are studied in two different environments: the canonical and microcanonical ensembles. A comparison is made to thermodynamic properties of the bulk. It is shown that consistent and reproducible results on nanoclusters can only be obtained in the canonical ensemble. Nanoclusters in the microcanonical ensemble are trapped systems, and inconsistencies will be found if thermodynamic formalism is applied. An analytical model is given for the energy dependence of the phase space volume of nanoclusters, which allows the prediction of both dynamical and thermodynamical properties.

Keywords: nanocluster; thermodynamics; heat capacity; dynamics; canonical ensemble; microcanonical ensemble

1. INTRODUCTION

Nanoclusters are small finite systems consisting of from several atoms to several thousands of atoms. These systems are interesting because their physical, optical and electronic characteristics are strongly size dependent. Often changing the size by only one atom can significantly alter the physical-chemical properties of the system. Many new periodic tables can thus be envisioned in which differently-sized clusters of the same material become the new elements. Potential applications are enormous, ranging from devices in nano-electronics and nano-optics [1] to applications in medicine and materials. A revolution in technology based on nanoclusters is at the point of fruition, predicted to be on a scale as large, or larger, than either the industrial or digital revolutions.

However, before the engineering of new designs and devices can be contemplated, complete physical characterization of nanoclusters is needed. Of particular importance is the understanding of the dynamic and thermodynamic behavior of these systems in different environments [2,3,4,5,6].

The present article studies the dynamics and thermodynamics of nanoclusters in two different environments, the microcanonical and canonical ensembles. It will become clear that the choice of the environment, or ensemble, in which the thermodynamic properties of the nanocluster are to be determined is crucial, since they generally lead to different, and therefore inconsistent, results. This is because in the microcanonical ensemble, nanoclusters inevitably become trapped at low energy in only a sub-region of the energetically-available phase space, and so, the system never arrives at a true thermodynamic equilibrium, independently of the amount of time allowed for convergence. Consistent and reproducible thermodynamic results cannot therefore be expected, and any results so obtained will depend on initial conditions or subsequent perturbation. However, under some circumstances, and under the premise that the initial conditions can be ascertained, dynamical information obtained in the microcanonical ensemble may be useful in describing the behavior of nanoclusters in a given practical application.

The work presented in this paper concerning the trapping by large energy barriers has relevance not only to nanoclusters, but to all "small" systems, in which the size of the system is small with respect to the range of the inter-particle forces. This occurs for any size of system if the inter-particle potentials decay with exponents smaller than the dimensionality of the embedding space, for example in the case of self-gravitating systems. In these cases, the time "relaxed" distributions will be very dependent on initial conditions [7]. Since these systems can never obtain thermodynamic equilibrium, the application of the thermodynamic formalism to these systems is unjustified and will lead to inconsistencies in the formalism, including the apparent violation of fundamental law and negative heat capacities [8,9].

The canonical ensemble is more suited to obtaining reproducible thermodynamic information, which is independent of the initial condition or subsequent perturbation in the time-relaxed state. This is because in the canonical ensemble, the nanoclusters' environment is a thermal reservoir, which allows the exchange of energy, permitting the cluster to pass energy barriers and thereby visit all of its available phase space, at least in principle. However, even in the canonical ensemble, caution must be exercised before assuming thermodynamic equilibrium. Energy barriers between different isomeric forms (between solid-solid or between solid and mixed solid-liquid forms) of the nanocluster can be very high. Surmounting them can take excessively large times that are impractical to be simulated in numerical experiments or observed in real experimental conditions.

This situation, whether in the microcanonical or canonical ensembles, has been conceptualized as broken ergodicity and has been a constant source of confusion, stimulating much heated debate within the nanocluster community [4,10,11,12,13].

Applying the equilibrium thermodynamic formalism to a system trapped out of equilibrium will lead to inconsistencies and paradoxes with the postulates and theorems of the thermodynamic formalism [14]. One of these inconsistencies has been the reporting of the so-called "negative heat capacity", resulting from the *a priori* presumption of the validity of the results of the thermodynamic formalism obtained in the microcanonical ensemble.

The basic ergodicity hypothesis of molecular simulation ensures that thermodynamic results are equal in the microcanonical and canonical ensembles if the simulation is ergodic and well converged. The difference between results obtained in the canonical and microcanonical ensembles leads some researchers to conjecture the existence of two thermodynamics for small systems [5,8]. Our stance regarding this controversy has been to emphasize that nanoclusters in the microcanonical ensemble are not systems that can reach thermodynamic equilibrium, and therefore, it is inconsistent to apply the equilibrium thermodynamic formalism to such systems [4]. An important conclusion that came out of this previous work was that, since all macroscopic results obtained on systems trapped out of equilibrium will depend on the initial conditions (which, for nanoclusters, still cannot be adequately controlled in the laboratory) or depend on subsequent perturbation, many published results on nanocluster thermodynamics should be carefully re-examined.

The focus of our present study of the thermodynamic properties of nanoclusters will be the caloric curve, and thus, we first describe the anatomy of this curve in detail in Section 2. In Section 3, we emphasize the differences to be expected for this curve between that obtained on material in the bulk and that obtained on nanoclusters. In Section 4, we mention the frequent, but false assumptions and caveats normally encountered in the literature on the thermodynamics of nanoclusters. In Section 5, we present a simple model for obtaining the energy dependence of the density of states of a nanocluster in true thermodynamic equilibrium. This density of states can only be correctly obtained in the canonical (or macrocanonical) ensemble, since trapping in the microcanonical ensemble is inevitable. In the same section, we consider a modification of our model for the thermodynamics that allows us to include energy barriers and, thus, to simulate what would be expected in the microcanonical ensemble for particular initial conditions. This allows initial condition-dependent dynamical information to be obtained for the system. A comparison is then made of our results with caloric curves and heat capacities obtained using molecular dynamics at constant energy (the microcanonical ensemble). Section 6 presents a discussion and conclusions.

2. THE ANATOMY OF A CALORIC CURVE

The caloric curve is the graph of the temperature of a system as a function of its internal energy. The temperature is defined as the inverse of the partial derivative of the entropy with respect to energy at constant particle number and constant volume: $T=(\partial S \partial E)-1V,N$. It is thus an inverse measure of how fast the entropy is increasing with respect to the energy; if the entropy increases slowly with energy, the temperature is high, and *vice versa*. At the melting energy of the cluster, the caloric curve may demonstrate an inflection, which is due to the sudden increase in the number of modes for storing energy as the cluster becomes liquid (atoms in the liquid can carry translational and rotational kinetic energy related to their diffusion within the volume). Figure 1 shows a typical caloric curve for a nanocluster demonstrating a phase transition at an energy of 7.5 eV.

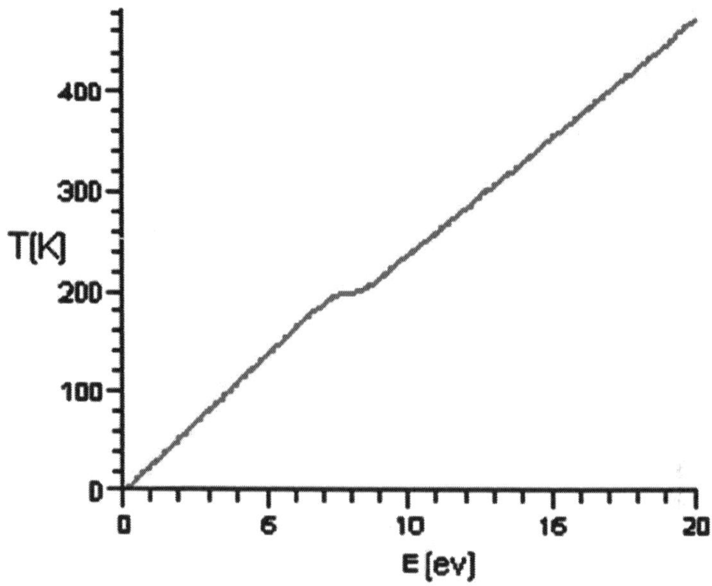

Figure 1. Example of a caloric curve, for a nanocluster.

The heat capacity is the inverse slope of the caloric curve, CV,N=(∂T∂E)−1V,N, and is thus an inverse measure of how fast the temperature is changing with energy; if the temperature increases slowly with energy, the heat capacity is high, and *vice versa*.

For a cluster of N atoms, which is neither rotating nor translating (angular and linear momentum fixed at zero), there are only $3N - 6$ normal vibrational degrees of freedom. Each vibrational degree of freedom contributes two modes for storing energy; one in the kinetic energy associated with vibration and one in the potential energy (these are the two quadratic terms that enter the Hamiltonian for the solid). Each of these contributes $1/2k_B$ to the heat capacity (assuming that the energy is high enough to excite all vibrational modes). The heat capacity of the solid is thus $(3N - 6)k_B$, the Dulong–Petit law, and the slope before the transition (where the system is solid) is therefore $1/(3N - 6)k_B$. This slope before the transition is practically the same for both the nanocluster and for the bulk. Note that the contribution to the slope of the caloric curve before the transition is the same for whatever symmetry (isomeric form) in which the cluster or bulk may be. This is simply because there are always only $3N - 6$ vibrational degrees of freedom. (It is convenient to point out, however, that, if at low energies, not all vibrational modes can be excited or if strong anharmonicities in the potential are relevant, then the slope of the caloric curve before the transition may, in general, differ from the Dulong–Petit law.) The energy intercept of the caloric curve can be different, however, depending on the binding energy of the particular symmetry.

Approaching in energy from below the solid-to-liquid transition, the slope of the caloric curve for a nanocluster decreases (the heat capacity increases) because new modes for storing energy come into play. These modes are related to the fact that the surface atoms are now relatively free to move over the surface of the underlying solid structure and, thus, can carry energy in their translational modes on the two-dimensional surface of the structure. The $3N - 6$ vibrational degrees of freedom remain even though they take on a lower frequency and generally become anharmonic. Each of these new translational modes contributes a further $1/2k_B$ to the heat capacity, and thus, the heat capacity increases as $(3N - 6 + i)k_B$, where i is the number of atoms in the liquid (the cluster may be a mixed solid and liquid).

The heat capacity can also be written in terms of the entropy as the second partial derivative of the entropy with respect to energy divided by the rate of change of entropy with energy (at constant volume and particle number),

$$C_{V,N}(E) = \left(\frac{\partial^2 S}{\partial E^2}\right)_{V,N} \Bigg/ \left(\frac{\partial S}{\partial E}\right)_{V,N} \tag{1}$$

It is also possible to write the heat capacity in terms of the phase space volume (or density of states) $\Gamma(E)$ using Equation (1) and the Boltzmann equation, $S(E) = k_B \ln\Gamma(E)$. This gives,

$$C_{V,N}(E) = \left[1 - \frac{\Gamma(E)\Gamma''(E)}{\Gamma'(E)^2}\right]^{-1} \tag{2}$$

The heat capacity thus depends not only on the phase space volume Γ, but also on its first and second derivatives with respect to energy, $\Gamma' \equiv \partial\Gamma/\partial E$ and $\Gamma'' \equiv \partial 2\Gamma/\partial E2$ (at constant volume V and particle number N).

From Equation (2), we note that for the heat capacity to go negative, it is required that $\Gamma\Gamma''/\Gamma' > 1$. Such a change in the phase space volume will be referred to as "abrupt" in the following. Borrowing from the language of cosmology, one could say that the heat capacity goes negative when the phase space inflation, defined as $\Gamma\Gamma''/\Gamma'^2$, becomes greater than unity. It can be ascertained that this will happen only if the phase space volume Γ increases faster than a power law or, more exactly, faster than a non-infinite sum of power laws, with energy [4]. However, quantization of the extensive variables prohibits the existence of such systems in nature (see [15] and the Discussion and Conclusions section of this paper).

3. BULK *VERSUS* NANOCLUSTER THERMODYNAMICS

The microstates contributing to the thermodynamics of bulk material (containing over 10^{23} atoms) are overwhelmingly determined by the atoms of the volume. Except for particular processes, such as, for example, the initiation of melting, surface atoms play very little part in the thermodynamics, simply because there are too few surface atoms as compared to the vast amount of atoms of the bulk. Although a given bulk material may have a number of possible symmetries (e.g., fcc, bcc, hcp, *etc.*) these symmetries are separated from one another in state space by very large energy barriers that can only be surmounted at the melting temperature (the phase transition). At the energy of melting for the bulk, in the microcanonical ensemble, many new modes for storing energy become simultaneously accessible, and the phase space increases instantaneously, leading to a singularity in the caloric curve, which takes the form of a delta function. This leads to a discontinuous increase in the entropy and is the description of a first order phase transition.

Below the transition energy in bulk materials, not all microstates have *a priori* equal probability because there exists other symmetries corresponding to phase space regions that are never sampled because of the enormously large energy barriers between symmetries. However, for all practical purposes, before the transition, the other symmetries of the bulk can be safely ignored, and we can consider the system as having obtained thermodynamic "equilibrium" while existing in only one particular symmetry. This holds for both the microcanonical and canonical ensembles.

Different symmetries of the same material below the melting transition are named differently (e.g., graphite and diamond), and the caloric curve and all other thermodynamic results for a given symmetry are always reproducible, because the initial condition is easily determined and the same from one experiment to the next.

The thermodynamics of nanoclusters is different from that of the bulk precisely because the number of surface atoms is greater than, or of the order of, the number of volume atoms. The surface atoms have lower coordination (fewer nearest neighbors) and are thus less bound to the rest of the cluster. This fact, coupled with the finite size of the system, often implies that the different symmetries available to the cluster are separated from each other by barriers that can be below the melting energy. Solid to solid transitions are thus allowed before the melting transition [16], a possibility generally not available to the bulk.

The finite size of the system also means that new, additional, symmetries (for example, the icosahedra and decahedral) are allowed for the nanocluster, while they are strictly forbidden for the bulk. Furthermore, because nanoclusters are very small systems, thermodynamic averages (macroscopic results) imply that measurement must be made on an ensemble of identical systems. It is currently impossible experimentally to be able to control the initial conditions, and so, the distributions of the clusters among their many allowed symmetries will depend very much on the non-equilibrium process used to form them.

Since it is not possible to control initial conditions and, furthermore, since each member of the ensemble can change symmetry even before the phase transition energy, the only way to obtain useful (reproducible) information concerning the behavior of nanoclusters is to assure that one has a true thermodynamic equilibrium distribution over all symmetries at all energies. This, because of trapping by barriers, can only be obtained by employing a canonical ensemble in which energy can be utilized from the thermal reservoir to surmount the barriers.

Thus, for a nanocluster, different symmetries can come into play at total energies below the solid to liquid transition, and this is due to a combination of factors involving the total binding energy and the configurational and vibrational entropies of the cluster, as well as the structure and strength of the bonds of the outer layers. The phase space volume of a cluster thus increases more gradually near the transition than does that of the bulk, and there exists no singularity, but rather, an extended region in energy over which the heat capacity gradually increases as the system gains new modes for storing energy. These modes for storing energy are related to the fact that the atoms in the liquid are no longer constrained to particular symmetries, but are relatively free to pass over barriers and, so, can carry kinetic energy of diffusion over the two-dimensional surface of the underlying solid part of the cluster.

However, it is not always true that the energy barriers between different symmetries are below the solid to liquid transition energy. In Figure 2, we show that the energy barriers (blue, online, or dashed lines, in print) between the icosahedral (red, online, or dark grey, in print) and decahedral (green, online, or light grey, in print) isomers for the nanocluster Na_{147} are large and, in fact, reach up to the energy (≈7.8 eV) of the transition region. Note, however, that in a truly equilibrium distribution, the microstates corresponding to the decahedral isomers should be accounted for at an energy of ≈0.8 eV, which is well below the barrier energy of ≈7.8 eV.

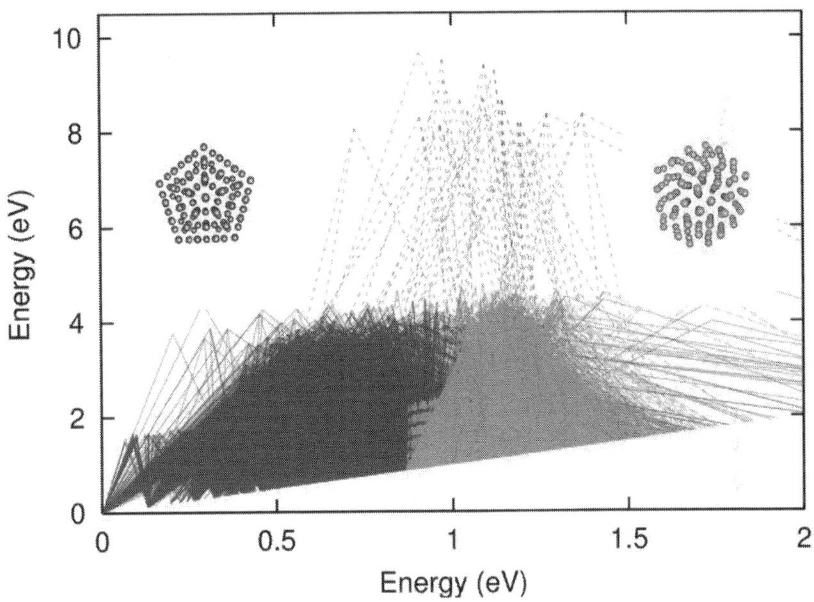

Figure 2. Plot showing the heights of the energy barriers (blue, online, or dashed lines, in print) between various isomers of the icosahedral (red, online, or dark grey, in print) and decahedral (green, online, or light grey, in print) isomers of Na_{147}. The red (dark grey, in print) lines mark the energy barriers between different isomers of icosahedral symmetry while the green (light grey, in print) lines mark the energy barriers between different isomers of decahedral symmetry. Data were obtained using a genetic algorithm to search the Gupta potential energy surface. Details are given in [17].

4. FREQUENT CAVEATS ENCOUNTERED IN THE ANALYSIS OF THE THERMODYNAMICS OF NANOCLUSTERS

As mentioned above, the facts that clusters can become easily trapped in different symmetries, that this trapping is a function of energy and that initial conditions cannot be adequately controlled are at the origin of the inconsistencies found in the application of the equilibrium thermodynamic formalism to determine the thermodynamic properties of atomic nanoclusters.

The energy barriers between different isomers of the same symmetry are usually small, while those between isomers of different symmetry are large and often reach into, or beyond, the transition region (see Figure 2). For sufficiently low energy in the microcanonical ensemble, each cluster of the ensemble, just as in the case for the bulk, is trapped in one particular symmetry, which depends on its initial condition. However, unlike in the case of the bulk, the cluster may leave that particular symmetry of its initial condition before reaching the transition energy. As a specific example, Figure 3demonstrates this "untrapping" of isomers before the transition for LJ_7 (a cluster of seven Lennard-Jones potential bound atoms). The figure shows the "caloric curves" obtained in the microcanonical ensemble (at constant energy) by starting the molecular dynamic simulations from the four different isomers of LJ_7, whose geometry is shown in the inset of the figure. From the figure, it can be ascertained that below an energy of 1.5 in units of ε (ε is the depth of the potential well for binary interaction), the cluster is trapped in the particular isomer defined by its initial condition. Hence, four distinct caloric curves are obtained on increasing the total energy, each corresponding to a different initial condition or isomer. Untrapping occurs at 1.5 ε, which is considerably lower than the transition energy of 3.5 ε.

For larger-sized clusters, the energy barriers between the different symmetries become higher and move closer in energy to the transition region. For larger systems, there is also the possibility of new microstates corresponding to part of the cluster in the solid phase and part of the same cluster in the liquid phase (referred to as phase separation in the literature). These states may include one, two, three, *etc.*, up to all *N* atoms in the liquid.

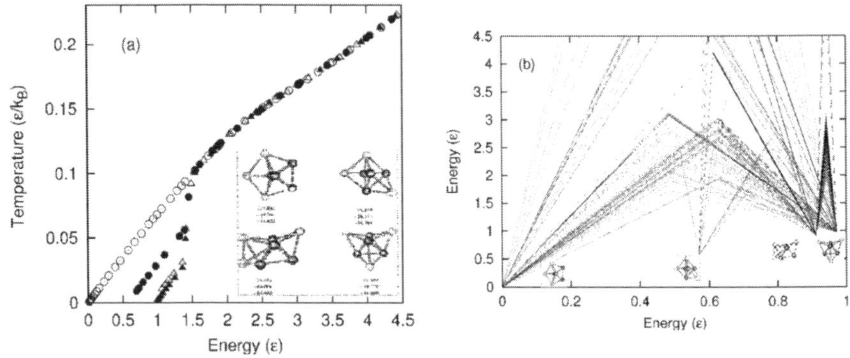

(a) (b)

Figure 3. (**a**) Four different caloric curves are obtained for LJ_7 by starting constant energy molecular dynamics simulations in the different isomers shown in the inset. This dependence on initial condition highlights the unavoidable trapping in the microcanonical ensemble. (**b**) Energy barriers between the different isomers obtained through a genetic algorithm search on the potential energy landscape, described in [17]. The barriers found in (b), the lowest at 1.5 ε, are in agreement with the results obtained in (a).

It has often been assumed that these mixed solid-liquid states do not contribute to the thermodynamics below the transition energy, as it was argued that an energy cost of a presumed interface between the solid and liquid had to be paid [6]. The nature of the interface, however, was never explained nor investigated in the subsequent literature. The only support for this hypothesis was the observation that only rarely were mixed solid liquid states found in Monte Carlo and molecular dynamics simulations of clusters. These mixed states were, therefore, hitherto, not given their proper consideration in the thermodynamic analysis of clusters.

Our interpretation, however, is that these states were rarely observed, not due to their high energy, but instead because of the high energy barriers separating them from other regions of the space of available states. The heights of the energy barriers separating these mixed solid-liquid states from the solid states depend not only on how many atoms are in the liquid, but also on the nature of the symmetry of the underlying solid structure. For a large enough nanocluster in the microcanonical ensemble, these mixed solid-liquid states will be hidden from the simulation or experiment until surmounting their particular energy barriers. For the case of the magic number clusters (corresponding to closed geometry shells of a given symmetry), the barriers separating the mixed states from the solid states may be higher than the barrier between all solid and all liquid states (even though the mixed states are available, but not accessible at a significantly lower energy than all of the liquid states). Therefore, in the microcanonical ensemble, increasing the energy of the cluster may lead to a direct jump from the all-solid (or nearly all solid) to the all-liquid (or nearly all liquid) states. In this case, the measured phase space inflation may become so large that the apparent heat capacity, as determined by Equation (2), may become negative. We emphasize, however, that what is being measured is not a true heat capacity, since the system is not in thermodynamic equilibrium in the microcanonical ensemble because of trapping. The mixed solid-liquid states are hidden by the large energy barriers, but must be accounted for in an equilibrium calculation. Exactly how all this comes about and leads to the erroneous determination of "negative heat capacity" will be made clear in Section 5 with the help of our model for the energy dependence of the phase space volume of a cluster.

As mentioned in the Introduction, a further indication of the inconsistency resulting from ignoring the mixed solid-liquid states and, thus, incorrectly assuming thermodynamic equilibrium is the difference in results obtained in the microcanonical and canonical ensembles. The ensemble defines the different environment, but it should not affect the sought after thermodynamic properties of the system under consideration. There is only one thermodynamic equilibrium state, and this state must be determined identically and independently of the ensemble used. Ignoring this fact has led researchers to consider the inconsistencies in the analysis as exotic thermodynamic attributes [18,19].

5. MODEL FOR THE PHASE SPACE VOLUME OF A NANOCLUSTER

We seek an analytical model that can accurately describe the true thermodynamics of nanoclusters. Preferably, such a model should allow for the introduction of energy barriers between different microstates of the system in such a way that we may also use the model to describe the spectrum of the dynamical behavior that would be observed for a trapped nanocluster in the microcanonical ensemble. Furthermore, we would like our model to be valid over a useful range of energies, including the phase transition from solid to liquid.

We begin by considering the possible microstates available to a nanocluster assuming only two phases, the solid and liquid. We seek the energy dependence of the phase space volume $\Gamma(E)$ for an ensemble of clusters in thermodynamic equilibrium. In practice, this could be obtained from the energy probability density $\rho(E,T_0)$ of a canonical ensemble of clusters in true thermodynamic equilibrium at temperature T_0;

$$\Gamma(E) = \rho(E,T_0) Z \exp(-E / k_B T_0) \tag{3}$$

with Z the partition function. At any temperature T_0, for each member of the ensemble, there may be one, two, three, *etc.*, up to N (N = number of atoms in the cluster) atoms in the liquid. Each of these possibilities contributes to $\rho(E,T_0)$ and thus to $\Gamma(E)$. Every one of these possibilities is a microstate that must be counted, weighted (in the canonical ensemble, the microstates have different probabilities [14]) and averaged over in the determination of the thermodynamics.

An alternative, but equivalent and more straightforward procedure is to assume no energy barriers and to develop the phase space volume in the microcanonical ensemble. We have presented this approach in detail in another publication [4], so here, we only give a summary of the procedure. We construct a model for the phase space volume of a nanocluster, including the change at the solid to liquid transition, by considering the contribution to the total phase space of each partial volume Γi constituted by i indistinguishable particles in the liquid phase in interaction with $N - i$ indistinguishable particles in the solid phase:

$$\Gamma_i(E) = \left(N!/(N-i)!i!\right)\Gamma_i^l(E)\Gamma_{N-i}^s(E) \tag{4}$$

where Γli(E) and ΓsN−i(E) are the liquid and solid phase space volumes. The combinatory factor gives the number of ways to distribute i particles from N identical particles between the liquid and solid phases. The total phase space volume is then obtained by summing over all possible partial volumes constituting all distributions of particles among the solid and liquid: Γ(E)=∑Ni=0Γi(E). Note that the microcanonical ensemble imposes no restrictions on how the energy E is distributed over the system.

The phase space volume for a classical system increases as a power law in the energy, where the power exponent is equal to one half the number of quadratic terms that enter the Hamiltonian for the system. Thus, a simple model of the energy dependence of the phase space volume, which includes the change in the number of degrees of freedom upon melting, for a cluster of N atoms (with total momentum and total angular momentum fixed at zero) is,

$$\Gamma(E) = c\sum_{i=0}^{N} \frac{N!}{(N-i)!i!}\left(\frac{E}{E^*}\right)^{6(N-i)/2-6} \cdot \left(\frac{E}{E^*}\right)^{(6+\lambda)i/2} \tag{5}$$

where c is a constant and E^* is the characteristic energy of the solid to liquid transition. In the harmonic approximation, E^* can be taken as the geometric mean of the normal mode vibrational frequencies v_i of the solid, E∗≈ℏ·∏3N−6i=1νi-------√3N−6. The first energy factor in Equation (5) corresponds to contributions to the phase space volume due to the solid part of the cluster, while the second corresponds to contributions due to the liquid part, with the extra λ modes for storing energy per atom. As mentioned above, for a cluster, there are nominally $\lambda = 2$ new modes related to the translational energy of the atoms moving over a two-dimensional surface. The different values of i correspond to the different grades of melting; for $i = 0$, the cluster is all solid; for $i = N$, the cluster is all liquid; for $i = N/2$, half of the atoms are in the solid part and half are in the liquid part of the cluster.

Since the thermodynamics is independent of an arbitrary constant multiplying the phase space volume (only changes in entropy are important), Equation (5) may be rewritten in the more condensed form,

$$\Gamma(E) = d\sum_{i=0}^{N} \frac{1}{(N-i)!i!}\left(\frac{E}{E^*}\right)^{3N+i-6} \tag{6}$$

Given this density of states, it is then straightforward to calculate the thermodynamics of the system using the Boltzmann equation, $S(E) = k_B \ln \Gamma(E)$. The temperature is then given by,

$$T(E) = \left(\frac{\partial S(E)}{\partial E}\right)^{-1}_{V,N}$$

and the specific heat (heat capacity per atom) by,

$$C_{V,N}(E) = \frac{1}{N}\left[1 - \frac{\Gamma(E)\Gamma''(E)}{\Gamma'(E)^2}\right]^{-1}$$

The caloric curve and the specific heat for the cluster Na_{147} obtained using the density of states as determined by Equation (6) are given by the solid line in Figure 4a,b, respectively. These curves represent the results that would be obtained for a cluster in thermodynamic equilibrium. In this case, all mixed microstates (different terms in Equation (6) corresponding to a different number of atoms in the liquid) are included. The specific heat shows a smooth increase from its value of $(3N - 6)/N$in units of k_B to its value of $(4N - 6)/N$. The transition is not the abrupt first order transition of the bulk, but of order N.

We may now ask what would happen if our measurements were performed in the microcanonical ensemble where they are affected by the energy barriers. This would correspond to the dynamics of a trapped system. If we assume the extreme case in which all mixed solid-liquid states have high barriers reaching into the transition region, then our system would go directly from the all-solid into the all-liquid state. To model the energy dependence of this phase space volume, it is necessary to remove all intermediate terms in Equation (6) corresponding to the mixed solid-liquid states. This leaves us with just two terms, that of the all-liquid and that of the all-solid state,

$$\Gamma(E) = d\left[\left(\frac{E}{E^*}\right)^{3N-6} + \left(\frac{E}{E^*}\right)^{4N-6}\right] \tag{7}$$

The corresponding "caloric curve" and "specific heat" curves are given by the dashed lines in Figure 4a,b, respectively. In this case, the transition is not smooth, but abrupt, and the measured specific heat becomes negative. This, as mentioned above, is not a thermodynamic result, since the measured phase space volume given by Equation (7) does not correspond to a system in equilibrium, but instead to a trapped system.

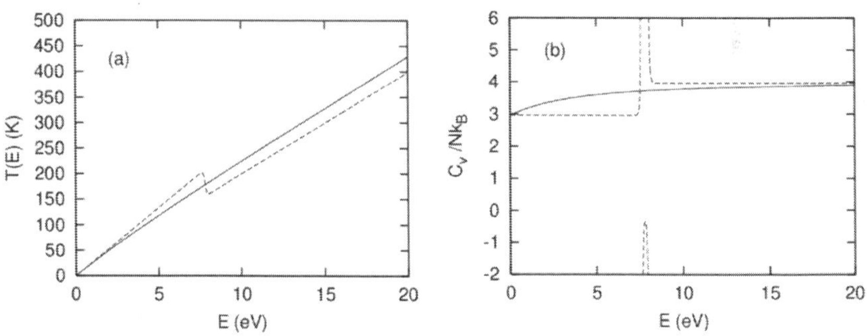

Figure 4. (**a**) Caloric curves for Na_{147} obtained with the full phase space volume, Equation (6) (solid line), and sampled non-ergodically for the extreme case of all the atoms going to the liquid simultaneously in energy, Equation (7) (dashed line); (**b**) the corresponding specific heats. Apparent negative specific heat is found for the non-ergodic case. Taken from [4].

In reality, different mixed solid-liquid states could have energy barriers of different heights. In fact, surface melting has been readily observed in simulations of core-shell nanoalloys [20,21,22,23]. A more correct description of the energy dependence of the measured phase space volume for such a trapped system is therefore not the extreme case as given by Equation (7), but one in which we have a different E^* for each mixed solid-liquid state. Our Equation (6) then becomes:

$$\Gamma(E) = d\sum_{i=0}^{N} \frac{1}{(N-i)!i!}\left(\frac{E}{E_i^*}\right)^{3N+i-6} \tag{8}$$

What E^*'s to use depend on the energetics of liquid over solid structures. Obtaining the E∗i from the potential model or experiment directly is not as difficult a task as it would seem at first sight. Given the experimental or simulation specific heat curves, we can fit the specific heat curve obtained by theory, using Equation (8) in Equation (2), by varying the E∗i. Figure 5shows the result of fitting the theoretical specific heat to the specific heat determined through molecular dynamic results at constant energy [24] minimizing the χ^2 of the fit by varying the E∗i's. The corresponding best fit barrier heights E∗i for the different mixed solid-liquid states with *i* atoms in the liquid are given in Figure 6.

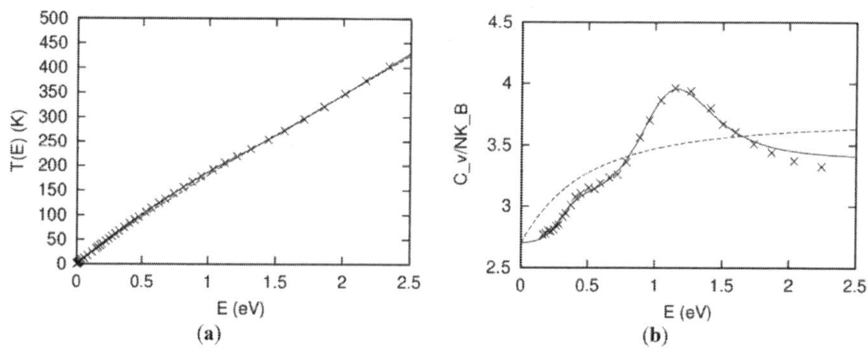

Figure 5. (**a**) Caloric curve for Na_{20} obtained from molecular dynamics runs at constant energy [24] (crosses) and using the model phase space volume Equation (6) (dashed curve). The solid curve is a fit using Equation (8) with the E∗i as given in Figure 6. (**b**) Corresponding specific heat curves.

Note that as suggested above, the all-solid to all-liquid barrier (E∗i, i = 20) is lower than the E∗i for many mixed solid-liquid states. It is this fact that leads to the false determination of negative heat capacity for larger-sized clusters (e.g., Na_{147}).

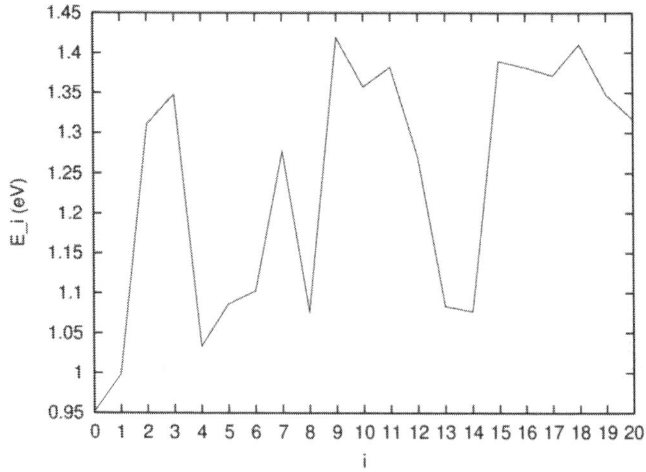

Figure 6. The energy barriers E∗i for the mixed solid-liquid states with i atoms in the liquid for Na_{20}. Obtained by fitting the model specific heat capacity, using the phase space volume of Equation (8), to molecular dynamics results at constant energy (seeFigure 5).

6. DISCUSSION AND CONCLUSIONS

We have studied the dynamics and thermodynamics of nanoclusters in two different environments, the canonical and the microcanonical ensembles. We have shown that equilibrium thermodynamic properties can only be obtained from the canonical ensemble and that nanoclusters in the microcanonical ensemble are trapped systems. Negative heat capacities and differences between results obtained in the canonical and microcanonical ensembles are simply inconsistencies that result from applying the equilibrium thermodynamic formalism to a cluster trapped out of equilibrium.

Nanoclusters differ in their statistical properties from the bulk in that many possible symmetries contribute to the accessible microstates and because energy barriers between solid and mixed solid-liquid states are variable in energy and often below the solid to liquid transition. This is due to the fact that surface atoms are important in the thermodynamics of nanoclusters, and these are proportionately greater in number and less tightly bound than in the bulk.

We have given a simple analytical model to describe the dynamics and thermodynamics of nanoclusters and have shown how this model can be used to understand the inconsistencies in the analysis, such as the determination of negative heat capacity and the differences obtained in the two different ensembles (canonical and microcanonical). We have also used our model to determine the barrier heights for the mixed solid-liquid states, a model that we suggest will be of great utility to the study of the dynamics of nanoclusters.

Finally, within the confines of the heated debate regarding the validity or not of negative heat capacity in nanoclusters, we are often confronted with toy statistical models, which apparently do not suffer from broken ergodicity in the microcanonical ensemble and, yet, demonstrate negative heat capacity (see, for example, [10,11,25,26]). However, in all such cases that have been presented to us so far, we have shown that these models do not represent physically-realistic systems that can reach thermodynamic equilibrium. The thermodynamic formalism can thus not be legitimately applied to these models.

The reason for the illegitimacy of applying the thermodynamic formalism to these toy statistical models has been given in [15]. There, we have shown that under the two basic axioms that must be satisfied in order to derive equilibrium thermodynamics from statistical mechanics, (i) the independence of the microstates and (ii) the *a priori* equal probability of the microstates, the density of states of a system cannot have any arbitrary dependence on the extensive variables of the system. In particular, we showed that in order to comply with these two basic axioms, the system must have a density of states $\Gamma(X)$ dependence on the extensive variable X, which obeys the following relation;

$$\left[\frac{\Gamma(X)\Gamma''(X)}{\Gamma'(X)^2}\right] < 1 \tag{9}$$

for any extensive variable and where the prime refers to the differentiation of the density of states with respect to the extensive variable. This can be seen to be just Relation (2) for assuring a positive heat capacity for the case of the extensive variable X being the energy E.

If Relation (9) is violated for any extensive variable X by a particular model, then the model violates one or more of the two basic axioms and, as such, is not treatable under the thermodynamic formalism. A careful analysis of Relation (9) shows that valid thermodynamic models must have a number of microstates that grows not faster than a power law or, more exactly, not faster than a non-infinite sum of power laws, with any extensive variable [15]. If one anyhow insists on applying thermodynamic formalism on these models that do not satisfy Relation (9), then inconsistencies in the analysis will result, one of them being the determination of "negative heat capacity".

Since any system trapped out of equilibrium will demonstrate macroscopic results that depend on the initial conditions, we suggest that all previous "thermodynamic" results on nanoclusters should be carefully re-examined.

ACKNOWLEDGMENTS

Karo Michaelian is grateful for the hospitality afforded by L.M. Gaggero of the Facultad de Ciencias, Universidad Autonoma del Estado de Morelos, Morelos, México, and for the financial support of DGAPA-UNAM Project IN-103113. I.S.-H. is grateful for the financial support of DGAPA-UNAM Project IN-113415.

REFERENCES

1. Andres, R.P.; Bein, T.; Dorogi, M.; Feng, S.; Henderson, J.I.; Kubiak, C.P.; Mahoney, W.; Osifchin, R.G.; Reifenberger, R. "Coulomb Staircase" at Room Temperature in a Self-Assembled Molecular Nanostructure. *Science* **1996**, *272*, 1323–1325.

2. Engelmann, Y.; Bogaerts, A.; Neyts, E.C. Thermodynamics at the nanoscale: Phase diagrams of nickel-carbon nanoclusters and equilibrium constants for phase transitions. *Nanoscale* **2014**, *6*, 11981–11987.

3. Jellinek, J.; Goldberg, A. On the temperature, equipartition, degrees of freedom, and finite size effects: Application to aluminum clusters. *J. Chem. Phys.* **2000**, *113*, 2570–2582.

4. Michaelian, K.; Santamaría-Holek, I. Critical Analysis of Negative Heat Capacity in nanoclusters. *Europhys. Lett.* **2007**,*79*, 43001.

5. Wales, D.J.; Berry, R.S. Coexistence in Finite Systems. *Phys. Rev. Lett.* **1994**, *73*, 2875–2878.

6. Wales, D.J.; Doye, J.P.K. Coexistence and phase separation in clusters: From the small to the not-so-small regime. *J. Chem. Phys.* **1995**, *103*, 3061–3070.

7. Levin, Y.; Pakter, R.; Rizzato, F.B. Collisionless relaxation in gravitational systems: From violent relaxation to gravothermal collapse. *Phys. Rev. E* **2008**, *78*, 021130.

8. Lynden-Bell, D.; Lynden-Bell, R.M. On the negative specific heat paradox. *Mon. Not. R. Astron. Soc.* **1977**, *181*, 405–419.

9. Lynden-Bell, D. Negative specific heat in astronomy, physics and chemistry. *Physica A* **1999**, *263*, 293–304.

10. Calvo, F.; Wales, D.J.; Doye, J.P.K.; Berry, R.S.; Labastie, P.; Schmidt, M. Comment on "Critical analysis of negative heat capacities in nanoclusters" by Michaelian, K. and Santamaría-Holek, I. *Europhys. Lett.* **2008**, *82*, 43003.

11. Lynden-Bell, D.; Lynden-Bell, R.M. Negative heat capacities do occur. Comment on "Critical analysis of negative heat capacities in nanoclusters" by Michaelian, K. and Santamaría-Holek, I. *Europhys. Lett.* **2008**, *82*, 43001.

12. Michaelian, K.; Santamaría-Holek, I. Reply to the Comment by D. Lynden-Bell and R.M. Lynden-Bell. *Europhys. Lett.* **2008**, *82*, 43002.

13. Michaelian, K.; Santamaría-Holek, I. Reply to the Comment by F. Calvo *et al. Europhys. Lett.* **2008**, *82*, 43004.

14. Callen, H.B. *Thermodynamics and an Introduction to Thermostatistics*, 2nd ed.; Wiley: New York, NY, USA, 1985.

15. Michaelian, K.; Santamaría-Holek, I.; Pérez-Madrid, A. Violation of Fundamental Thermodynamic Law or Invalid Microstate Densities? 2013, arXiv:1307.7736v1.

16. Cleveland, C.L.; Luedtke, W.D.; Landman, U. Melting of Gold Clusters: Icosahedral Precursors. *Phys. Rev. Lett.* **1998**,*81*, 2036.

17. Michaelian, K.; Taméz, A.; Garzón, I.L. Efficient statistical mapping of the energy surfaces of nanoclusters and molecules. *Chem. Phys. Lett.* **2003**, *370*, 654–660.

18. Helmer, M. Only iced tea in nanoland. *Nat. News* **2001**.

19. Schmidt, M.; Schmidt, M.; Kusche, R.; Hippler, T.; Donges, J.; Kronmüller, W.; von Issendorff, B.; Haberland, H. Negative Heat

Capacity for a Cluster of 147 Sodium Atoms. *Phys. Rev. Lett.* **2001**, *86*, 1191–1194.

20. Aguado, A.; López, J.M. Molecular dynamics simulations of the melting-like transition in $Li_{13}Na_{42}$ and $Na_{13}Cs_{42}$clusters. *Phys. Rev. B* **2005**, *71*, 075415.

21. Kuntová, Z.; Rossi, G.; Ferrando, R. Melting of core-shell Ag-Ni and Ag-Co nanoclusters studied via molecular dynamics simulations. *Phys. Rev. B* **2008**, *77*, 205431.

22. Sankaranarayanan, S.K.R.S.; Bhethanabotla, V.R.; Joseph, B. Molecular dynamics simulation study of the melting of Pd-Pt nanoclusters. *Phys. Rev. B* **2005**, *71*, 195415.

23. Van Hoof, T.; Hou, M. Structural and thermodynamic properties of AgCo nanoclusters. *Phys. Rev. B* **2005**, *72*, 115434.

24. Reyes-Nava, J.A.; Garzón, I.L.; Michaelian, K. Negative heat capacity of sodium clusters. *Phys. Rev. B* **2003**, *67*, 165401.

25. Bixon, M.; Jortner, J. Energetic and thermodynamic size effects in molecular clusters. *J. Chem. Phys.* **1989**, *91*, 1631–1642.

26. Carignano, M.A.; Gladich, I. Negative heat capacity of small systems in the microcanonical ensemble. *Europhys. Lett.***2010**, *90*, 63001.

3

CHAPTER

Thermodynamics and Irreversibility: From Some Paradoxes to the Efficiency of Effective Engines

Olivier Serret

ESIM Engineer—60 rue de la Marne, Cugnaux, France

ABSTRACT

The traditional thermodynamic theory explains the reversible phenomena quite well, except that reversible phenomena are rare or even impossible in practice. Here the purpose is to propose an explanation valid for reversible and also irreversible phenomena, irreversibility being common or realistic. It previously exposed points tricky to grasp, as the sign of the work exchange, the adiabatic expansion in vacuum (free expansion) or the transfer of heat between two bodies at the same temperature (isothermal transfer). After having slightly modified the concepts of heat transfer (each body produces heat according to its own temperature) and work (distinguishing external pressure from internal pressure), the previous points are more easily explained. At last, an engine efficiency in case of irreversible transfer is proposed. This paper is focused on the form of thermodynamics, on "explanations"; it does not question on "results" (except the irreversible free expansion of 1845...) which remain unchanged.

Keywords: Heat Transfer, External and Internal Pressure Work, Internal Energy, Reversible and Irreversible, Joule's Law and Joule's Experiments, Adiabatic and Free Expansion, Clapeyron Diagram, Carnot Cycle, Engine Efficiency

1. INTRODUCTION

Thermodynamics traditionally [1] [2] deals of relations between the thermal and the mechanical phenomena. It is a science that was born in the nineteenth century and was codified by scholars as Sadi CARNOT or James Prescott JOULE. Since then, little major changes have been made, on form and content. It is on this first point that presentation takes place in this essay. Indeed, some statements remain difficult to understand, even if current "explanations" do not taint the veracity of the "results".

In the first part, we will check some statements tricky to understand: a mathematical equality considered as a physical principle, an expansion without any work, a negative work represented by a positive area, a transfer of heat between two bodies at the same temperature, and a cycle where heat seems to go from cold to hot! In the second part, we will change the forms of the heat transfer and of the work and consequently of the internal energy. Then in the third part, we will apply these new forms to the first statements which were tricky to understand. Explanations look then more understandable, especially for irreversible phenomena. At last, an efficiency ratio based on differences of temperatures is proposed for irreversible cycles.

2. SOME STATEMENTS TRICKY TO UNDERSTAND

2.1. The "Principle Zero" of Thermodynamics

Its name or more precisely its numbering—"zero"—is unusual. The reason is historical. The first principles having already been laid, it has nevertheless been considered necessary to add another one prior to the demonstrations. But more than its numbering, it is its content which is surprising as a "principle". Recall that the principle zero states that if A is in (thermal) equilibrium with B and B with C, then A is with C. This is surprising because that looks like more to a mathematical property (a transitive law, an equivalence relation or an axiom as "two quantities equal to a third are equal") than to a physical principle. In other branches of physics, it is not stated as a "principle" that if A has the same mass as B and B as C, then A has the same mass as C. This zero principle asks a question: does not it reflect a certain difficulty in characterizing thermal equilibrium?

2.2. Free Expansion and Joule Experiment

2.2.1. Historical

In 1806, Louis GAY-LUSSAC was the first to experiment expansion in vacuum. Measuring the temperature directly on the air with an alcohol thermometer, he found no variation of temperature [3] .

In 1845, James JOULE did the same experiment, except that it measured the temperature of the surrounding water; he found no variation of temperature either [4] .

In 1865, Gustave-Adolphe HIRN did the same experiment, measuring the temperature through a variation of pressure. He found a light variation of temperature: −0.2°C [5] . See Figure 1.

Since there, the experiment has been done again with carbon dioxide CO_2 (which is more sensitive than air): it has been noted a decrease of temperature of −0.3°C [6] [7] .

And today, it is mainly the usual industrial way to cool gases [8] [9] .

And yet History has retained the result of the Joule's experiment with no variation of temperature!

2.2.2. Explanation

Let us remind that an ideal gas is a gas that obeys the ideal gas law:

$$P \cdot V = n \cdot R \cdot T \tag{1}$$

with P: internal pressure of the gas; V: volume of gas; n: amount of substance of the gaz (in moles); R: gaz constant (8.314 $J \cdot K^{-1} \cdot mol^{-1}$); T: absolute temperature (in Kelvin).

This is a good approximation for gases in usual conditions (when the pressure is relatively low, less than 10 atm.) like the atmospheric pressure. It is often used to describe cycles, compressions and expansions.

There are two laws of Joule: the first one called Joule-Gay Lussac law and the second one called Joule- Thomson law. The first law states that for ideal gas the internal energy depends only on the temperature, the

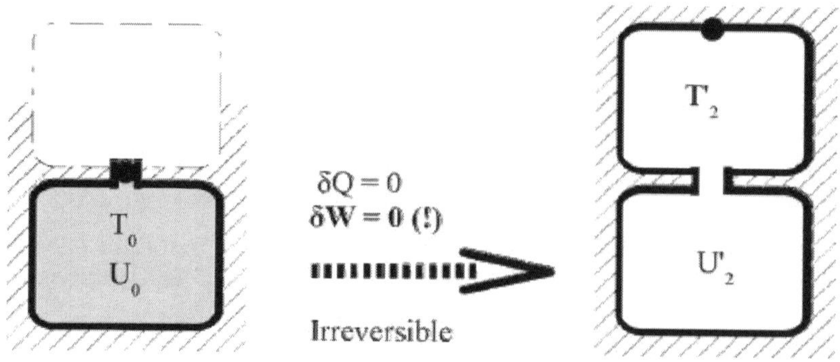

Figure 1. The so-called Joule's experiment.

second one that the enthalpy depends only on its temperature. To check it, these scholars set up an experiment showing that "the gas temperature would remain constant" (which is quite different to the Joule's law: U depends only on its temperature) with the expansion, due to the formula

$$dU = \delta Q + \delta W \tag{2}$$

The reasoning is as follow: because there is no heat transfer, and because there is no work ($P_{vacuum} = 0$ so without external pressure, there would not be pressure work), there is no change of internal energy U and so we should not have change of temperature.

This is why it is still traditionally explained that if the gas cools in experiments, it is that in sudden expansion the gas is far to be ideal [the gas would not obey to Equation (1)]: this is an indirect recognition that traditional theory is not in agreement with experimental results...

2.2.3. A Surprising Result

What a surprising result: a thermodynamic system which keeps constant its internal energy U and its temperature T, and which does not receive any heat or work but submits an irreversible change! It is the result found by Pr. Joule in 1845.

Why is it qualified here of surprising result? Because it is unusual to get:

no work when the volume changes; it would be surprising that the border line, which can be the black top of Figure 1, moves without any work! It would be at variance with the first principle of thermodynamics where perpetual motion machines of the first kind are impossible...

· a constant temperature

$T_2' = T_0$ (3)

when in the reversible and in the other irreversible cases, the temperature decreases [10] with expansion (without heat transfer); please check on Figure 2.

2.3. Work Representation by Its Area

In mathematics, the area is calculated by integrating a function. If the function $f_{math}(x)$ is positive so its integral, represented by an area, is positive: $[F_{math}] > 0$ [11] . See Figure 3 where $f_{math}(x)$ is $P_{ext}(V)$ and $[F_{math}]$ is $[W_{math}]$.

In thermodynamics, work is conventionally counted positively when received and negatively when provided by the system. The infinitesimal work exchanged by the system is denoted δW.

$$\delta W = -P_{ext}\mathrm{d}V \tag{4}$$

The pressure P_{ext} being always positive, when V_2 is higher than V_1 (that is, when x_2 higher than x_1), the thermodynamics sum W is then negative: $W < 0$. How explain that in the [P-V] Clapeyron diagram particularly suitable for representation of work, the negative work W is paradoxically represented by a positive area W_{math}?

2.4. Carnot Cycle

The theorizing of thermodynamics by physicists did not precede but followed or accompanied the development by engineers of thermal engines as the steam engine. They had already realized that the efficiency η (work

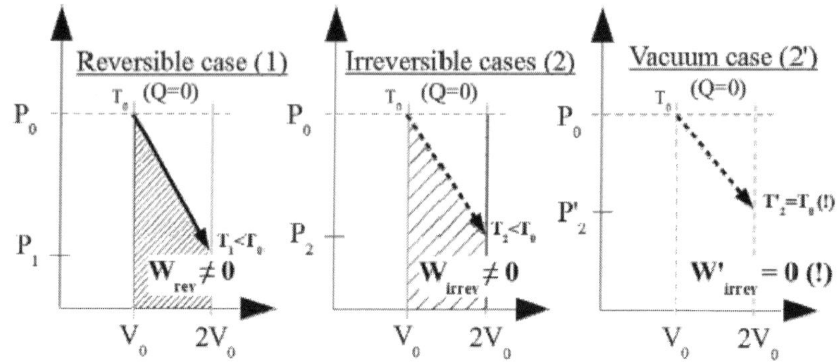

Figure 2. Work and temperature evolution with expansion.

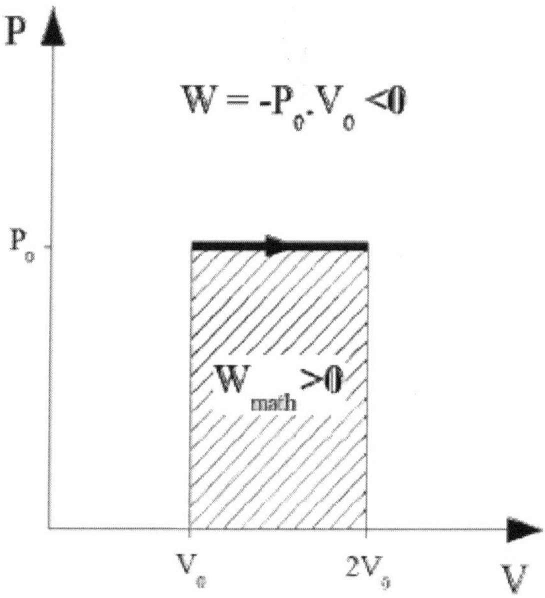

Figure 3. Representation of positive mathematical area.

done/energy transmitted by the hot source) seemed limited by an unsurpassable value related to the temperatures of the sources.

$$\eta \leq 1-\frac{T_{\text{Cold}}}{T_{\text{Hot}}} \tag{5}$$

Was it an impression or a physical principle? Sadi Carnot, who thought the heat was a fluid called caloric, described a clever cycle (not representative of an existing machine) which reaches the performance limit. Let us consider a transformation of this cycle, the isothermal transformation. SeeFigure 4.

An isothermal transformation is a transformation that takes place at constant temperature, in contact with a thermostat. Take for example a piston air-filled at 20° Celsius, in an external medium like water also at 20° Celcius. The air volume is adapted to the pressure according to the ideal gas law. This system is in stable equilibrium; it does not change and will not change spontaneously. And there is no reason that for two bodies at the same temperature, heat flows in one direction rather than another. In the case of perfect equality between fluid temperature and thermostat temperature: there is no transfer of heat.

$$Q_{\text{Hot}} = Q_{\text{Cold}} = 0 \tag{6}$$

And more generally for isothermal transformation

$$T_{\text{fluid}} = T_{\text{thermostat}} \rightarrow Q = 0 \tag{7}$$

This cycle cannot work as a thermal engine.

The efficiency η of the Carnot cycle is defined as the ratio of work supplied of the heat input, (i.e. by energy conservation), dividing heat balance by heat input:

$$\eta = \frac{Q_{\text{Hot}} - Q_{\text{Cold}}}{Q_{\text{Hot}}} \tag{8}$$

Hence [cf Equation (6)]

$$\eta = \frac{0}{0} \tag{9}$$

η is not calculable. Rigorously, the Carnot efficiency (based on isothermal conditions) is not defined!

2.5. Carnot Engine

The thermal cycle of a motor is traditionally represented as follows in Figure 5.

In motor mode, the heat from the hot source is partly converted in form of work, the balance going to the cold source.

And the thermal cycle of a refrigerating unit is traditionally represented as follows in Figure 6.

In refrigeration mode, the heat from the cold source goes to the hot source and the work too. How is it possible that the heat goes from cold to hot?

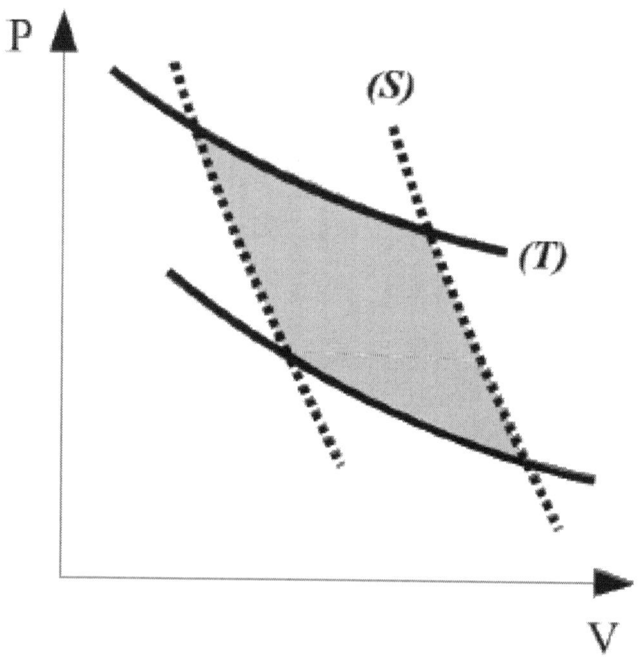

Figure 4. Carnot cycle.

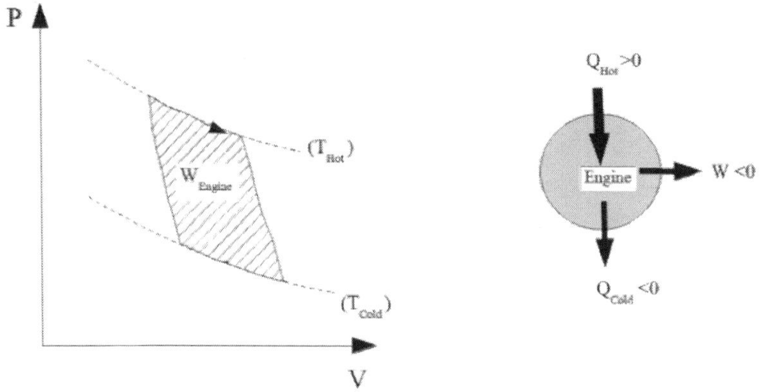

Figure 5. Traditional representations of motor cycle.

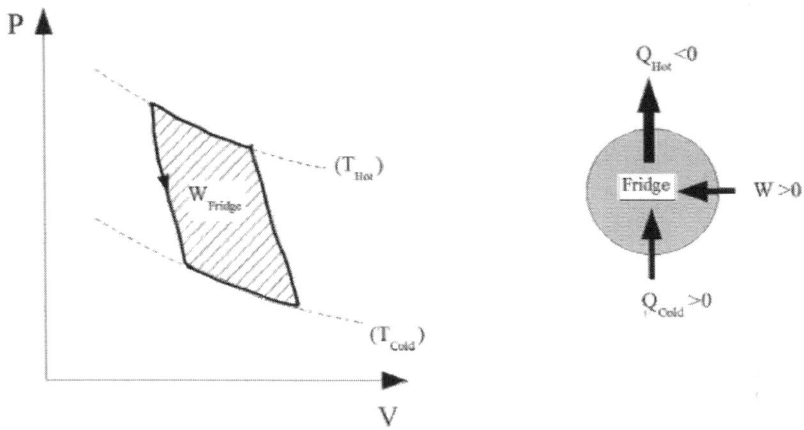

Figure 6. Traditional representations of cooling cycle.

3. STATEMENT OF ASSUMPTIONS

The above points will be discussed again after the following assumptions:

3.1. Internal Energy U

Beforehand, what is the internal energy U? "The internal energy of given state cannot be directly measured. (···) Though it is a macroscopic quantity, internal energy can be explained in microscopic terms by two theoretical virtual components. One is the microscopic kinetic energy (···). The other is the potential energy (···). There is no simple universal relation between these quantities of microscopic energy and the quantities of energy gained or lost by the system in work, heat, or matter transfer." [12] . So the link between macroscopic and virtual microscopic components is:

$$U = \sum(E_K) + \sum(E_P) \tag{10}$$

with E_K for kinetic energy of particles and E_P for potential energy

And the main property of U is: "The internal energy is a state function of the system, because its value depends only of the current state of the system and not on the path taken or processes undergone to prepare it." [8] . This property does not change here.

3.2. Emitted Heat Proportional to Temperature

The first new assumption is that all bodies emit heat, and this heat δQ is proportional to the temperature T of the body over a period dt:

$$\delta Q_{\text{emitted}} = s \cdot A \cdot T \cdot \mathrm{d}t \tag{11}$$

with:

- s coefficient, unit in [W/(K·m²)]
- A area of the body, in [m²]
- T temperature of the body surface, in [K]
- dt time length of the thermal exchange, in [s]

Remark: if entropy S in [J/K] has the properties of a status function, s is only a coefficient in [W/(K·m²)]. And like C_p, s is empirical and should be dependent for example on the temperature.

3.3. Work of Internal Pressure

According to the fundamental principle of dynamics, acceleration undergone by a body is proportional to the net strength it received. The piston is moved because at a time that internal strength was different from the external strength, i.e. the internal pressure was different from the external pressure. See Figure 7.

It is considered here (please read arguments in Appendix) the internal strength or internal pressure, and therefore their work according to the formula:

$$\delta W_{\text{int}} = F_{\text{int}} \cdot \mathrm{d}l \tag{12}$$

or

$$\delta W_{\text{int}} = P_{\text{int}} \cdot \mathrm{d}V \tag{13}$$

Remark: The fundamental principle of dynamics indicates that if internal pressure equals constantly and exactly to external pressure, then there is no movement (for a body initially at rest): the reversible movement where constantly and exactly $P_{\text{int}} = P_{\text{ext}}$ is impossible. The reversible movement is a limit but an unrealistic case.

3.4. Variation of Internal Energy Function of Heat Transfer and Internal Pressure Work

The property to calculate the "variation" of U is traditionally given by $\mathrm{d}U = \delta Q + \delta W$ [cf Equation (2)] with $\delta W = P_{\text{ext}} \cdot (-\mathrm{d}V_{\text{int}})$ [cf Equation (4)] which gives.

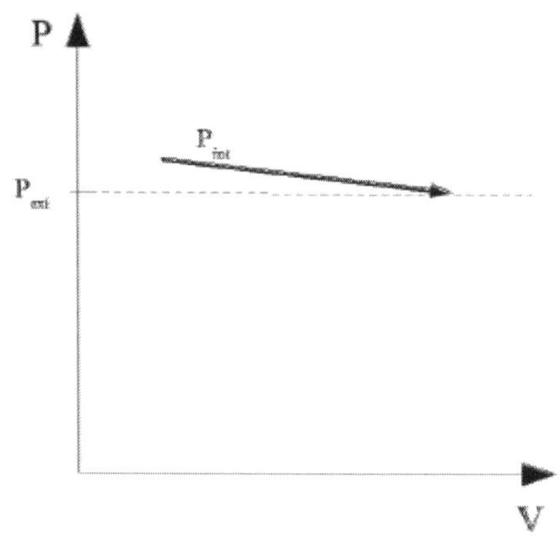

Figure 7. P_{int} was different from P_{ext}.

$$dU = \delta Q - P_{ext} \cdot dV \tag{14}$$

Note than the variation of the "internal" energy would depend from the "external" pressure.

In this essay let us rather have the internal energy to depend from the internal pressure:

$$dU = \delta Q - \delta W_{int} \tag{15}$$

or

$$dU + \delta W_{int} = \delta Q \tag{15bis}$$

with $\delta W_{int} = P_{int} \cdot dV_{int}$ [cf Equation (13)] which means

$$dU = \delta Q - P_{int} \cdot dV \tag{16}$$

3.5. Comments

- P_{ext} has only been changed into P_{int}, there is no major change in the expression of $|\delta W|$ and dU; no major change means that P_{int} has in practice the same magnitude than P_{ext} (except of course in the particular case of expansion in vacuum).
- When $P_{int} = P_{ext}$, then dU according to Equation (14) is equal to dU according to Equation (16):

$$\delta Q + P_{ext} \cdot (-dV) = \delta Q - P_{int} \cdot dV \tag{17}$$

- For reversible paths, $P_{int} = P_{ext}$, so there is no difference in the calculation of dU.
- For irreversible paths, we will get a difference in the calculation of $|\delta W|$. This difference of work calculation will change the "predicted" reached point in the Clapeyron diagram; for example, shall we reach the temperature T_1 or T_2? See Figure 8.

- Then, to calculate the difference of internal energy from the initial point (T_0), we will choose a "reversible" path, and we have demonstrated herebefore there was no difference in the calculation of dU_{rev}, and so in the calculation of the U status [we will find either $U(T_1)$ or $U(T_2)$].
- It confirms the U value of such defined point depends only of the current state of the system and not on the path taken or processes undergone to prepare it.
- For ideal gas, the property that the macroscopic internal energy U depending only on the temperature (the Joule's law) remains unchanged.
- Due to Equation (10) and Equation (15bis), the link between the macroscopic value U and the virtual microscopic terms could be changed.

To sum up into a chart (Chart 1).

4. CONSEQUENCES

In function of these hypotheses, points of Paragraph 2 are now discussed again.

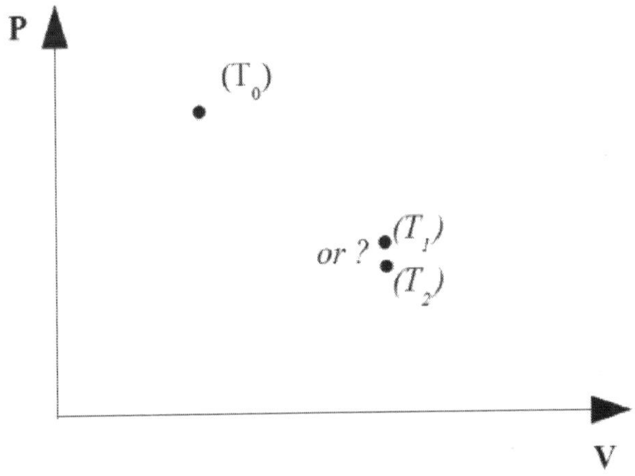

Figure 8. Change of the predicted reached point.

4.1. Thermal Balance and Heat Transfer

4.1.1. Concept

Suppose two identical bodies, except for temperature, in a fully insulated enclosure: all the heat emitted from one body A is received by the other body B (and vice versa). According to the first hypothesis of Equation (11) where $\delta Q_{\text{emitted}}$ is proportional to temperature, at every moment (seeFigure 9):

- the body A at the temperature T_A emits the heat $\delta Q_{\text{emitted}/A}$ and receives from B the heat $\delta Q_{\text{emitted}/B}$
- the body B at the temperature T_B emits the heat $\delta Q_{\text{emitted}/B}$ and receives from A the heat $\delta Q_{\text{emitted}/A}$
- For A, the balance between the emitted heat and the received heat is the heat $\delta Q_{\text{balance}/A}$:

$$\delta Q_{\text{balance}/A} = \delta Q_{\text{emitted}/B} - \delta Q_{\text{emitted}/A} \tag{18}$$

- For B, the balance between the emitted heat and the received heat is the heat $\delta Q_{\text{balance}/B}$:

$$\delta Q_{\text{balance}/B} = \delta Q_{\text{emitted}/A} - \delta Q_{\text{emitted}/B} \tag{19}$$

This heat $\delta Q_{\text{balance}}$ is nothing but the exchanged heat δQ in the traditional theory.

4.1.2. Generalization

- We can note in the case of this single two bodies that

$$\delta Q_{\text{balance}/B} = -\delta Q_{\text{balance}/A} \tag{20}$$

This is in agreement with energy conservation, what is lost from one side is gained from the other side.

- If for example $T_A > T_B$, then according to Equation (11) on emitted heat

$$\delta Q_{\text{emitted}/A} > \delta Q_{\text{emitted}/B} \tag{21}$$

and so

$$\delta Q_{\text{balance}/A} < 0 \tag{22}$$

The (hot) body A emits more heat than it receives.

By the same reasoning

Chart 1. Similarities and differences about internal energy.

	Theory	*Traditional*	Proposed
	Determination	*Cannot be directly measured*	Cannot be directly measured
U	For ideal gas	*U function of T only*	U function of T only
	Link macro/micro	$U=\sum(E_K)+\sum(E_P)$	?
	Property	$dU=\delta Q-P_{ext}\cdot dV$	$dU=\delta Q-P_{int}\cdot dV$
ΔU	Reversible	$\Delta U=Q-P\cdot\Delta V$	$\Delta U=Q-P\cdot\Delta V$
	Irreversible	$U(T_1)-U(T_0)=Q-\int P_{ext}\cdot dv$	$U(T_1)-U(T_0)=Q-\int P_{int}\cdot dv$

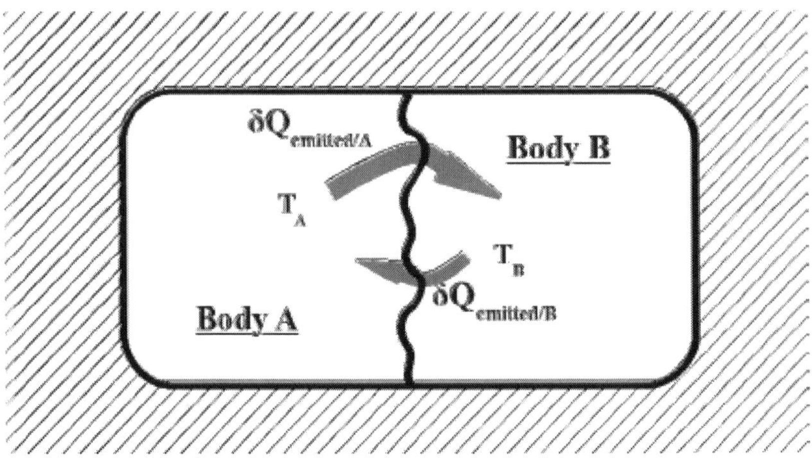

Figure 9. Heat transfer.

$$\delta Q_{\text{balance}/B} > 0 \tag{23}$$

The (cold) body B receives more heat than it emits.

Thus, the difference in heat is transferred from A to B, or in other words:

The heat "difference" goes from hot to cold.

Result: That prohibits the perpetual motion machine of the second kind where heat is directly removed from a body: for there to be heat transfer we need a heat differential that goes from hot to cold.

4.1.3. Case of Equilibrium

When $T_A = T_B$, then

$$\delta Q_{\text{balance}/A} = -\delta Q_{\text{balance}/B} = 0 \tag{24}$$

When heat balance is nil, the bodies remain in thermal equilibrium. But that does not mean the heat exchanged is non-existent, which means the exchange is balanced. That is why it is distinguished. So we will distinguish the "balance heat" $\delta Q_{\text{balance}/A}(=0)$ from the "emitted heat" $\delta Q_{\text{emitted}/A}(\neq 0)$.

Using this argument on three bodies, mathematics proved that two temperatures equal to a third one are equal, and so the three bodies are in stable equilibrium; it would not be longer a thermodynamical principle, it would be a mathematical property.

4.1.4. Extension

For radiative exchange, according to Stéfan-Boltzmann's law, the emittance M of a black body is:

$$M = \sigma T^4 \tag{25}$$

with $\sigma = 5.67 \times 10^{-8}\ \left[\text{W} \cdot \text{m}^{-2} \cdot \text{K}^{-4}\right]$.

Let us have the hypothesis where the variable s would be equal to kT^3:

$$s = kT^3 \tag{26}$$

Note in this hypothesis the variable s increases as a function of temperature.

For a given area,

$$\delta Q_{\text{balance}/A} = \delta Q_{\text{emitted}/B} - \delta Q_{\text{emitted}/A} = kT_B^4 \mathrm{d}t - kT_A^4 \mathrm{d}t \quad (27)$$

$$\delta Q_{\text{balance}/A} = \left(T_B - T_A\right) k \left(T_B^3 + T_B^2 T_A + T_B T_A^2 + T_A^3\right) \mathrm{d}t \quad (28)$$

For T_A and T_B of the same order of magnitude, let us have ε_T such as

$$T_A = T_B + \varepsilon_T \quad ,(29)$$

ε_T being small compared to T_A and T_B.

$$\delta Q_{\text{balance}/A} = \varepsilon_T \cdot k \cdot \left[4T_B^3 + \varepsilon_T \left(6T_B^2 + 4T_B \varepsilon_T + \varepsilon_T^2\right)\right] \mathrm{d}t \quad (30)$$

ε_T being small compared to T_B,

$$\delta Q_{\text{balance}/A} \approx \varepsilon_T \cdot 4k \cdot T_B^3 \cdot \mathrm{d}t \quad (31)$$

and with

$$C = 4k \cdot T_B^3 \cdot \mathrm{d}t \quad (32)$$

and $\varepsilon_T = T_A - T_B$ according to Equation (23):

$$\delta Q_{\text{balance}/A} \approx C\left(T_B - T_A\right) \quad (33)$$

On a low temperature range and for a given duration t, $\delta Q_{\text{balance}/A}$ would be proportional to $(T_B - T_A)$ for radiative exchanges.

4.2. Joule Experiment and Adiabatic Expansion

4.2.1. The Joule's Law and the Selected Experiments

First we have to distinguish

- the Joule's law: the internal energy depends only of the internal temperature for an ideal gas, from
- the so-called experiment of Joule: the adiabatic expansion in vacuum. It is a specific experiment because it is done against vacuum, and so it is irreversible. Let us remind experiments and industrial practice indicate that without heat transfer $(\delta Q = 0)$ the gases cool by relaxing (at temperatures lower than the Mariotte temperature and at low pressure).

4.2.2. Usual Explanations

In the case of the expansion against vacuum, there is no heat exchange or external work exchange. Thus, according to Equation (14) $\mathrm{d}U = 0$, so U = Constant, and therefore (see Joule's law) T = Constant. As this is in contradiction with the experimental results where the temperature is decreasing, it is said that the gases do not follow the ideal gas law [13] ! Equation (1) of ideal gas would be invalid in these cases. Why would not rather Equation (2) of internal energy and so Equation (14) be invalid for irreversible transformations? See Figure 10 on incompatible assumptions.

Another explanation is to assert absolute vacuum does not exist (even in the intergalactic space, there is one particle per m³), there is always some molecules and thus a (tiny) level of external pressure ζ, and therefore work of external forces:

$$\delta W_{ext} = -\zeta \cdot \mathrm{d}V_{int} \tag{34}$$

$$\mathrm{d}V_{int} > 0 \text{ so } \delta W_{ext} \leq 0 \text{ and } \mathrm{d}U \leq 0 \tag{35}$$

In this case, the gas should hardly cool (whereas in practice, the gas really cools).

4.2.3. Proposed Explanation

According to Equation (16): $\mathrm{d}U = \delta Q - P_{int} \cdot \mathrm{d}V$. The internal pressure of the gas being never nil, the internal work is not nil, $\mathrm{d}V_{int} > 0$, and $\delta Q = 0$, therefore $\mathrm{d}U < 0$: placing itself under the usual conditions (lower temperature than Mariotte temperature and so to a pressure where the molecules do not interact), the ideal gas must cool itself by relaxing in accordance with the present hypothesis, and in accordance with experiments (without making necessarily a perfect vacuum).

A more precise calculation will be to use the formula from the Appendix:

$$\mathrm{d}W_{\mathrm{int.}} = \left(P_{\mathrm{int.}} + P_{\mathrm{ext.}}\right)/2 \cdot \mathrm{d}V_{\mathrm{int}} \cdot \tag{36}$$

Because $\delta Q = 0$ and $P_{ext} = 0$,

$$\mathrm{d}U = -P_{\mathrm{int}}/2 \cdot \mathrm{d}V_{\mathrm{int}}, \tag{37}$$

$$\mathrm{d}U < 0 \text{ so } \mathrm{d}T < 0, \tag{38}$$

The experimental results are in agreement with the Joule's law: with the expansion work the temperature is decreasing; and so there is a loss of internal energy (gained by the expansion work).

Conversely, in case of compression, the temperature increases and the internal energy consequently.

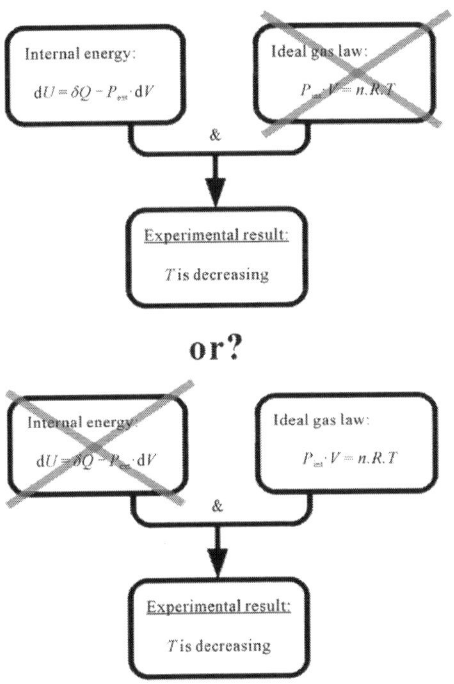

Figure 10. Incompatible assumptions.

4.3. The Area of Internal Pressure Work

During an expansion the volume increases, and so according to Equation (13) where $\delta W_{int} = P_{int} \cdot dV_{int}$: the internal work is positive $(W_{int} > 0)$.

When the abscise increases for a positive function, the mathematical integral represented by its area is positive $(W_{math} > 0)$. See Figure 11.

The same reasoning can be applied to the volume restriction and to the negative integral.

There is total correlation between the thermo dynamical work of the internal pressure W_{int} and the mathematical graphic representation W_{math} in the Clapeyron diagram in [P-V].

4.4. Engine Cycle

"A transformation is reversible if it can return to its initial position by a series of infinitesimal changes. It is only a thought experiment: the actual transformations are all irreversible" [14] .

The Figure 6 represents the idealized but incorrect case where heat is transferred whereas the internal fluid is at the same temperature as the hot source, and then at the same temperature as the cold source. For there to be heat transfer, "in practice, there must be some difference in temperature between the machine and the hot and cold sources" [15] . For the refrigerating machine, the fluid must be compressed to a temperature above the hot source, and then expanded to a temperature below the cold source for it to work. The Figure 6 should be reviewed as follows inFigure 12.

In refrigeration mode, heat of the cold source goes towards the colder fluid. Then the cycle fluid is compressed by the work done up to a temperature above the heat source for the heat dissipation of the fluid. And to close the cycle, the fluid is relaxed and cooled itself to return to the initial conditions.

The Figure 5 of the motor cycle are also idealized and incorrect, as there is no transfer of heat if the fluid is at the same temperature as the hot source, and then as the cold source. The Figure 5should be reviewed as follows in Figure 13.

In motor mode, heat from the heat source goes to the cooler fluid, the fluid expansion provides work (and to complete the cycle, the fluid transfers the heat balance to the cold source).

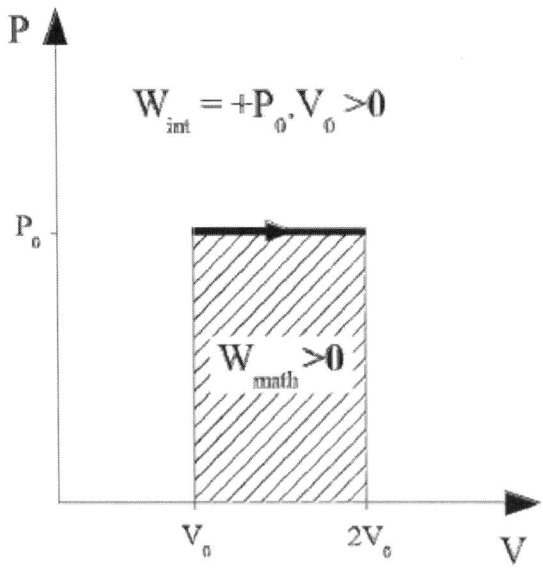

Figure 11. Representation of internal pressure work.

4.5. Motor Efficiency

4.5.1. Motor Efficiency Formula

And Figure 4 would have to be reviewed as follows in Figure 14.

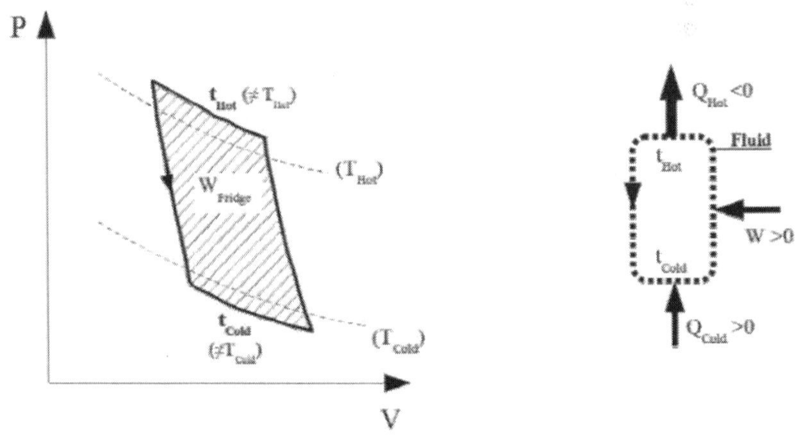

Figure 12. Proposed representation of cooling cycle.

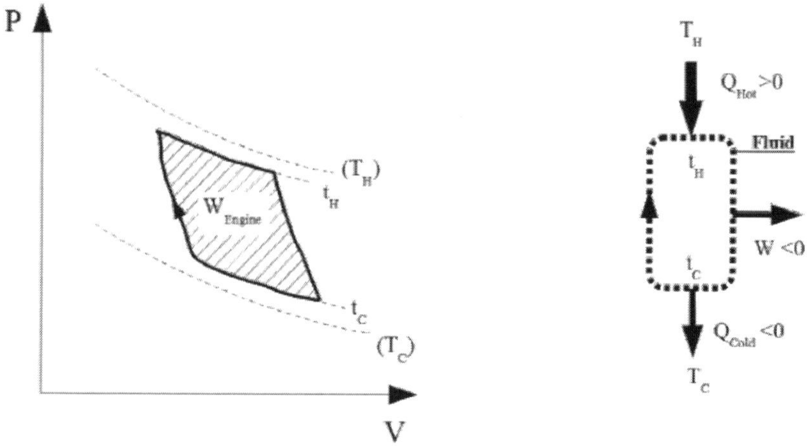

Figure 13. Proposed representation of motor cycle.

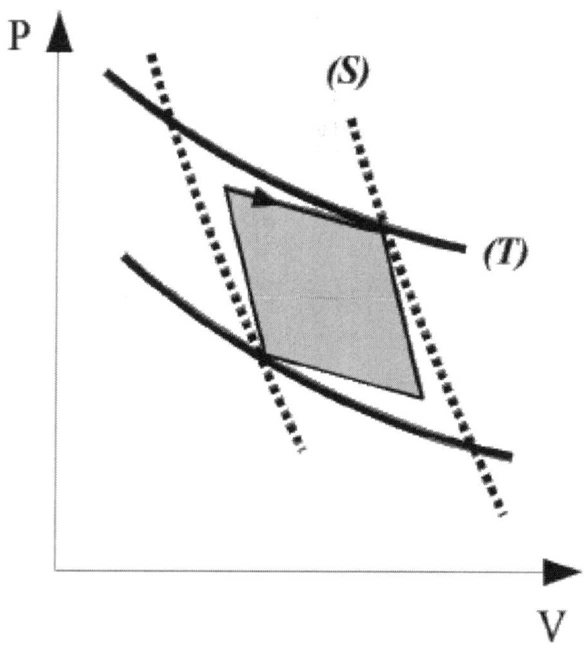

Figure 14. Proposed Carnot engine.

If this cycle is more realistic, it will generate complicated equations to calculate variable temperatures. That is why it is suggested here the following simplification, knowing that t_H and t_C are approximately (not exactly) constant. Let us characterize it with equations:

The heat transferred to the (relative) hot fluid from the Hot source is (cf Equations (11) and (18)):

$$Q_{\text{balance/hot fluid}} = s_H \cdot (T_H - t_H) \cdot t \text{ for a given area} \tag{39}$$

Note: the Hot source and the (relative) hot fluid being substantially at the same temperature, we consider firstly that they have substantially the same s_H (idem for s_C).

For writing formulas reason, let us simplify the expression: $Q_{\text{balance/hot fluid}}$ into Q_H (and $Q_{\text{balance/cold fluid}}$ into Q_C).

$$Q_H = s_H \cdot (T_H - t_H) \cdot t \text{ for a given area} \tag{40}$$

$$\text{As } T_H > t_H \text{ then } Q_H > 0 \tag{41}$$

And the heat transferred from the (relative) cold fluid to the Cold source is [cf Equation (11) and Equation (19)]:

$$Q_C = s_C \cdot (T_C - t_C) \cdot t \text{ for a given area} \tag{42}$$

$$\text{As } T_C < t_C \text{ then } Q_C < 0 \tag{43}$$

Preliminary remark: according that we adhere or not to the hypothesis expressed by Equation (15) taking into account internal work, demonstration hereafter does not change, it is why it is denoted by $|W|$, which can represent as well $|W_{\text{int}}|$ as $|W_{\text{ext}}|$.

For a motor cycle:

$$Q_H = |W| + |Q_C| \tag{44}$$

Motor efficiency η is:

$$\eta = \frac{|W|}{Q_H} \tag{45}$$

or according to Equation (44):

$$\eta = 1 - \frac{|Q_C|}{Q_H} \tag{46}$$

and using equalities (40) and (42):

$$\eta = 1 - \frac{s_C}{s_H} \cdot \frac{t_C - T_C}{T_H - t_H} \tag{47}$$

On the same range of temperature, $s_C \cong s_H$, so

$$\eta \cong 1 - \frac{t_C - T_C}{T_H - t_H} \tag{48}$$

which, for effective engine, is more accurate than the usual limit inequality $\eta \leq 1 - \frac{T_C}{T_H}$ [cf Equation (5)].

4.5.2. Theoretical Example

To get a better understanding, let us apply to a theoretical example, the irreversible Carnot engine, as described in Figure 15.

The effective efficiency is according to definition Equation (45):

$$\eta_{\text{eff.}} = \frac{|W|}{Q_H} = \frac{100}{67} = 60\% \tag{49}$$

If we use the efficiency of irreversible cycle based on temperatures as seen Equation (48):

$$\eta_{\text{irrev.}} \cong 1 - \frac{t_C - T_C}{T_H - t_H} \cong 1 - \frac{310 - 300}{800 - 775} \cong 60\% \tag{50}$$

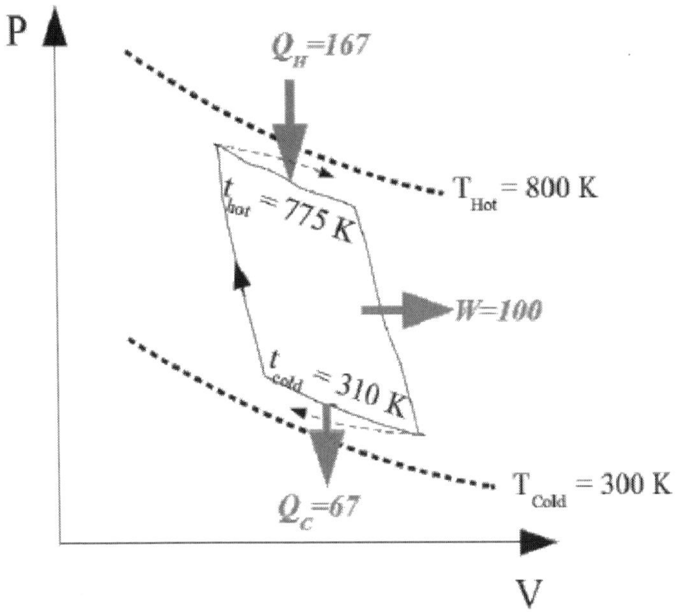

Figure 15. Irreversible Carnot engine.

which has the same value than the effective efficiency,.

Thus the Carnot efficiency based on sources temperatures is, according to Equation (5):

$$\eta \leq 1 - \frac{T_{\text{Cold}}}{T_{\text{Hot}}} = 1 - \frac{300}{800} = 62\% \tag{51}$$

which is slightly higher than the effective efficiency.

We can of course find the efficiency of 60% if we consider the cycle temperatures instead of the sources temperatures. The trouble is Carnot "engine" with isothermal evolution is unrealistic as explained §2.4. So, let us check another simple example.

4.5.3. Application

Then we will apply this property of the efficiency to a non-biphasic cycle: the very simple but more realistic elevator engine given by Pr. Richard Taillet [16] . Figure 16 and Chart 2 sum up the results:

In this non-biphasic cycle, the net work is −45 kJ for a heat transfer of (122 + 868 =) 990 kJ. The cycle effective efficiency is very low, (45/990 =) 5%.

When we apply the Carnot efficiency, with a 746 K hot temperature and a 300 K cold temperature, the maximum efficiency is according to Equation (5):

$$\eta_{\text{Carnot max.}} \leq 1-\frac{T_C}{T_H}=1-\frac{300}{746}=60\% \tag{52}$$

The maximum efficiency (60%) based on sources temperatures is very far of this cycle efficiency (5%).

To apply the new efficiency formula, we strictly should have used the instantaneous temperatures and integrate them, which would be very difficult here. So, only to give an order of magnitude, we will take in this example an average temperature of ((373 + 746)/2 =) 560 K for the hot side, and an average temperature of ((600 + 300)/2 =) 450 K for the cold side. An irreversible efficiency should be closed to

$$\eta_{\text{irrev.}} \cong 1-\frac{t_C-T_C}{T_H-t_H}=1-\frac{450-300}{746-560}=19\% \tag{53}$$

An irreversible efficiency (19%) based on differences of temperatures is closer to the effective efficiency (5%).

Previous calculation (19%) was done with a very approximative but simple average temperature only to explain. It has also been taken final fluid temperatures equal to the source temperature, which would mean a perfect heat exchanger. In practice, we will get for example 4 to 5 K between the final fluid temperature and the

Chart 2. Cycle results.

source temperature: $\Delta\theta$. Let us use the more accurate logarithmic difference of temperature (Δt_{ln}) used with heat exchangers:

$$\Delta t_{ln}=\frac{(T+\Delta\theta-t)-\Delta\theta}{\ln\left((T+\Delta\theta-t)/\Delta\theta\right)} \tag{54}$$

which would give

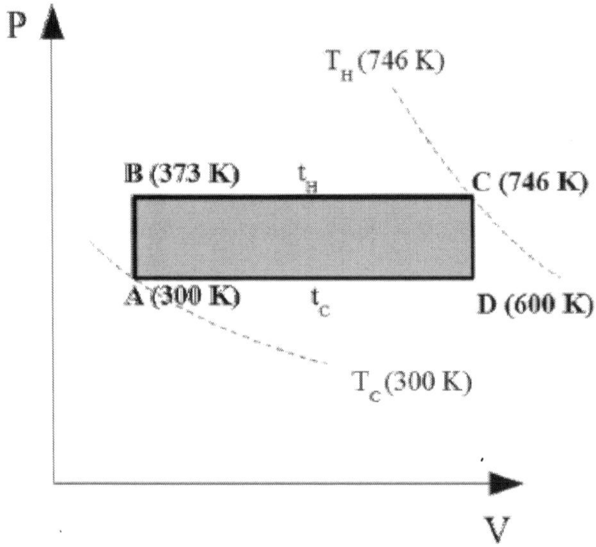

Figure 16. Non-biphasic motor cycle.

$$T_H - t_H = \Delta t_{ln/C} = \frac{(746+5-300)-5}{\ln((746+5-300)/5)} = 99\ \mathrm{K} \tag{55}$$

and, because $\Delta\theta$ is in practice a bit lower on the cold side than on the hot side

$$t_C - T_C = \Delta t_{ln/C} = \frac{(746-(300-4))-4}{\ln((746-(300-4))/4)} = 94\ \mathrm{K} \tag{56}$$

so

$$\eta_{\mathrm{irrev.}} \cong 1 - \frac{t_C - T_C}{T_H - t_H} = 1 - \frac{94}{99} = 5\% \tag{57}$$

which is in good correlation with the 5% of the effective efficiency.

5. CONCLUSION AND SUMMARY

Slightly modifying the concepts of heat transfer and work, we have given different explanations, and hopefully clearer on:

- the thermal balance and the heat transfer direction;
- the effects of expansion in the so-called experiments of Joule;
- the direct reading of the work in the Clapeyron diagram;

Chart 3. Comparison chart.

- a more accurate use of diagram of thermal cycles, especially for irreversible transformations in the motor cycle and the refrigeration cycle;
- a new property for irreversible engine efficiency: $\eta_{\text{irrev.}} \cong 1 - \frac{t_C - T_C}{T_H - t_H}$.

You will find a chart (Chart 3) to sum up the proposed formulas vs. the traditional formulas.

Without questioning the experimental results and the reversible predictions, the purpose of this essay is contributing to evolve the presentation of the thermodynamics traditional explanations, taking better account of Irreversibility.

ACKNOWLEDGEMENTS

Olivier Serret I would like to thank Daniel Mandineau for his teaching of traditional thermodynamics.

REFERENCES

1. Balian, R., CEA (2013) La longue élaboration du concept d'énergie. http://www.academie-sciences.fr/activite/hds/textes /evol_Balian1.pdf
2. Diu, B. (2001) Quatre questions sur la thermody namique. http://www.canal-

u.tv/video/science_en_cours/quatre_questions_sur_la_thermodynamique_posees_a_bernard_diu.19

3. ENS Cachan (1987) Gaz réels dilués, 140,141. http://www.physique.ens-cachan.fr/pagregp/enseignement /elec/corrigecompophy1987.pdf

4. Rowlinson, J.S. (2009) James Joule, William Thomson and the Concept of a Perfect Gas.http://rsnr.royalsocietypublishing.org/content/64/1/43.full, extract: In 1845 Joule had tried to verify this assumption, which Thomson always called "Mayer's hypothesis", by expanding a cylinder of compressed gas into an empty cylinder but, as he later admitted, he failed because the heat capacity of his vessels was too large compared with that of the gas. We now know that for air at atmospheric pressure and temperature the difference between the two terms on the right of this equation is only about 3 parts per thousand of either of them.

5. (1976) Détente, Larousse, http://www.larousse.fr/archives/grande-encyclopedie/page/4197 extract: The question arose whether the expansion of a gas in a vacuum, that is to say without external work, involving a change in temperature of the gas, assumed thermally insulated (expansion to constant internal energy); LJ Gay-Lussac then JP Joule have concluded in the negative, but their experiences were not very precise; GA Hirn (1865) brought to light a slight cooling.

6. Dr. P. Puzo from Institut National de Physique Nucléaire (2012) Description des fluides réels, 127. https://users.lal.in2p3.fr/puzo/thermo/ch6_thermo.pdf

7. Joule Expansion,Wikipedia. http://en.wikipedia.org/wiki/Joule_expansion

8. (2009) Gas Expansion Temperature Drop. http://www.oilngasseparator.info/gas-production-facility/hydrates/gas-expansion-temperature-drop.html

9. Rapin, P., Jacquard, P. and Desmons, J. (2011) Technologie des Installations Frigorifiques, Edition Dunod, 492. Extract: The obtained cooling [of air] is in the order of 0.25°C for a pressure drop of 1 bar.

10. Queyrel, J.-L. and Mesplede, J. (1996) Précis de Physique No. 6: Thermodynamique Prépas, Edition Bréal, 149.

11. Trizac, E. (1999) Cours de Thermodynamique. http://www.lptms.u-psud.fr/membres/trizac/Ens/ Livre_thermo.pdf
12. Internal Energy, Wikipedia. http://en.wikipedia.org/wiki/Internal_energy
13. Perez, J.-P. (2001) Détente de Joule et Gay-Lussac d'un gaz de Clausius.http://culturediff.pagesperso-orange.fr/jpp/ouvrage6/Thermo09-07.pdf
14. Serres, M. and Farouki, N. (1997) Le Trésor-Dictionnaire des Sciences. Flammarion, 855.
15. Gicquel, R., Mines de Paris Tech, Cycle de Carnot, p.4/9. l
16. Taillet, R., Université de Savoie (2012) Cours de thermodynamique.

4

CHAPTER

Second Law of Thermodynamics Formalism Applied to Finite Duration through Cycles of Living Dissipative Systems

Jorge Antonio Montemayor-Aldrete[1], Pablo Ugalde-Vélez[2], Marcelo Del Castillo-Mussot[1], Gerardo Jorge Vázquez[1], Ernesto Federico Montemayor-Varela[3]

[1]*Departamento de Estado Sólido, Instituto de Física, Universidad Nacional Autónoma de México, México, D.F., México*

[2]*Departamento de Materiales (CBI), Universidad Autónoma Metropolitana Azcapotzalco, México, D.F., México*

[3]*Instituto Nacional de Perinatología, México, D.F., México*

ABSTRACT

A simple and general theory to describe basic irreversible thermodynamic aspects of aging in all dissipative living is presented. Any dissipative system during its operation continuously loses efficiency by the production of structural or functional defects because of the second law of thermodynamics. This continuous loss of efficiency occurs on all the dissipative systems through the realization of specific functional cycles, leading to a maximum action principle of any system involving the Planck's constant during their total dissipative operation. We applied our theory to the calculation of men and women lifespans from basal metabolic rate per unit weight and to the calculation of a new aging parameter per cycle of some human organs or physiological functions. All microscopic theory of the aging of living beings should be consistent with the second law of the thermodynamics. In other words, the operation of the biological self-organized structures only implies a delay in which the dissipative biological systems outside of equilibrium approach inexorably to the thermodynamic equilibrium obeying the second law of the thermodynamics.

Keywords: Aging, Functional Damage, Entropy Production Rate, Dissipative Structures, Dissipative Cycles

1. INTRODUCTION

From Schrödinger's classic work published in 1944 [1] it became clear that life, and not only the not-living systems like cyclic thermal machines, obeys the first and second laws of thermodynamics. Relative to the human body he writes, "Energy is needed to replace not only the mechanical energy of our bodily exertions, but also the heat we continually give off to the environment. That our body gives off heat is not accidental, but essential. This is precisely the manner in which we dispose of the surplus entropy we continually produce in our physical life process, and seems to suggest that the higher temperature of the warm-blooded animal includes the advantage of enabling it to get rid of its entropy at a quicker rate so that it can afford a more intense life process". The basic same ideas were expressed in a different paragraph using similar expressions: "organization of living organisms is maintained by extracting 'order' from the environment" and he also asks: "How would we express in terms of the statistical theory the marvelous faculty of a living organism, by which it delays the decay into thermodynamical equilibrium (death)?" Finally, he introduced the idea of an "aperiodic crystal", and the microcode of which Schrödinger spoke have become the DNA and the genetic code of today. It is well known that Watson and Crick, co-discoverers of the structure of DNA, each respectively acknowledged the mentioned book as a source of inspiration for their initial researches.

The next step of the theoretical conception of living systems or non-living dissipative systems in a thermodynamical scheme was due to the Prigogine work in 1945, when the theorem of minimum entropy production applicable to non-equilibrium stationary states is published [2] . According to Prigogine this theorem gives a clear explanation of the analogy relating the stability of equilibrium thermodynamical states and the stability of biological systems, such as that expressed in the concept of "homeostasis" proposed by Claude Bernard. The notion of "dissipative structure" was elaborated by him in 1967 [3] , and in 1971, he concluded: "The formation of dissipative coherent structures, for sufficient far-from-equilibrium conditions, appears as a quite unexpected manifestation of 'disorder to order' processes, far from equilibrium, but conforming to the second law of thermodynamics" [4] . In the words, Professor Stig Claesson [5] states that: "Prigogine has been particularly captivated by the problem of explaining how ordered structures-biological systems, for example-can develop from disorder... which display order in both time and space, and which are stable to perturbations. Prigogine has called these systems dissipative systems, because they are formed and maintained by the dissipative processes which take place because of the exchange of energy between the system and its environment, and because they disappear if that exchange ceases... The method which Prigogine has used to study the stability of the dissipative structures to perturbations is of very great general interest. It makes it possible to study the most varied problems, such as city traffic problems, the stability of insect communities, the development of ordered biological structures... to mention but a few examples".

Additionally, to the previous information, it is an almost universal experimental fact that stationary states corresponding to dissipative structures do not last forever, despite being stable against fluctuations. Gradually, stationary states start to monotonically decline and eventually stop working to reach the thermodynamic equilibrium. For non-living dissipative systems, these processes could be defined as a process of intrinsic progressive loss of functionality, which conducts to ceasing of working properly, and eventually to failure [6] . For the living dissipative systems, this process of intrinsic progressive loss of organic function is reflected at the population level as an increase in the likelihood of death and a decline in the production of offspring [7] [8] .

Given the generality of the finite duration of stationary states in dissipative structures within irreversible thermodynamic processes, the following steps in a thermodynamical scheme, it is obligated. Let us try to use the second law of thermodynamics to describe the following facts. Section 2 contains a thermodynamical background necessary for our model. In Section 3, we present some empirical thermodynamical observations of living systems and present our very general theoretical formalism of an average dissipation rate theory taking into account the cyclic nature of living beings. In Section 4, we analyze in our formalism data comparing lifespans of human beings (females and males) and different human organs and physiological functions. Section 5 is devoted to conclusions.

2. THERMODYNAMICAL BACKGROUND

First, let us make a brief synthesis of relevant general thermodynamical results.

The first law reads,

$$\mathrm{d}U = T\mathrm{d}S - p\mathrm{d}V \tag{1}$$

where the internal energy of the system is U, T is temperature in Kelvin scale, S the entropy, p the pressure acting on the homogeneous system and V the volume. The second law defines entropy as

$$\mathrm{d}S = \frac{\mathrm{d}Q_{rev}}{T} \tag{2}$$

with Q_{rev} as the reversible heat flow.

Also we know [9] [10] that the rate of entropy production denoted by $\frac{\mathrm{d}_i S}{\mathrm{d}t}$, also can be expressed as a function of generalized fluxes J_i and generalized forces X_i:

$$\dot{S} = \frac{\mathrm{d}_i S}{\mathrm{d}t} \equiv \sum_i J_i X_i \tag{3}$$

From this equation, it is possible to define the Raleigh's dissipation function [10], Φ_R as

$$\Phi_R \equiv T\dot{S}_i \tag{4}$$

This represents the heat generation inside a system, and could be experimentally measured. For biological dissipative structures far from thermodynamic equilibrium, it is convenient to write Equation (4) per unit volume or per unit weight;

$$\varnothing_R = T\dot{s}_i \tag{5}$$

Prigogine [9] [11], by starting from Equation (3), shows that

$$\frac{\mathrm{d}_x P}{\mathrm{d}t} \leq 0 \tag{6}$$

where $P \equiv \frac{\mathrm{d}_i S}{\mathrm{d}t}$ and the symbolism given by $\frac{\mathrm{d}_x P}{\mathrm{d}t}$ denotes the time variation in P due to a variation on the forces. This result is valid provided the system is subject to time-independent boundary conditions, and it is the extension of the minimum entropy-production property to the non-linear domain of irreversible processes.

However, in contrast to inequality for linear irreversible thermodynamics $\left(\frac{\mathrm{d}P}{\mathrm{d}t} \leq 0\right)$ inequality Equation (6) does not imply the stability of the steady-state, primarily because in the general case $\mathrm{d}_x P$ is not the differential of a state function.

3. AVERAGE DISSIPATION RATE THEORY

3.1. General Empirical Facts

It has been experimentally observed long time ago that during the time development of living systems, the dissipative Raleigh function per unit volume $\varnothing_R$ grows rapidly with the beginning (after birth) of the living system, then later it attains a maximum value, and after that it decreases until a steady-state value is reached. However, this steady-state does not last forever, and eventually the system fails or dies. See for instance Figure 1 for the human normal basal metabolic rate at different ages for each sex.

The general schematic behavior of self-organized living dissipative structures is shown in Figure 2, where the specific (per unit volume) production rate of entropy versus time is presented. In more detail, the curve shows four chronological stages: 1) An initial transient stage, where $\dot{S}$ grows in a more or less exponential way; after some time, the production rate of entropy reaches its maximum value and becomes decreasing. In this stage, the organism is getting bigger and more developed through time than before; 2) Then we have a stationary or stea- dy-state stage, where $\dot{S}$ is approximately constant, which corresponds to a dynamical dissipative equilibrium state of the system. In this stage, the specimen already arrived to adulthood, and aging occurs in a slow rate; 3) "Aging in the mature stage" is characterized by an approximate linear decrease of with time, corresponding to an average linear decreasing of physiological response of all the organs or subsystems of the living being; 4) Finally, it occurs a stage corresponding to catastrophic failure or destruction of the biological dissipative structure.

In our formalism, the initial time is the instant when the "egg" is formed, which depends on the type of living being. For instance, in a mammal, we consider that time starts with the fetus development, and for the birds, the process depends on the heat received by the egg in the incubation process.

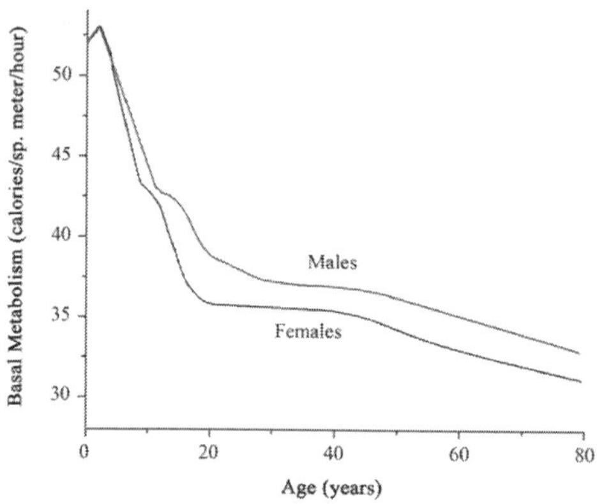

Figure 1. Normal basal metabolic rates at different ages for each sex [12] .

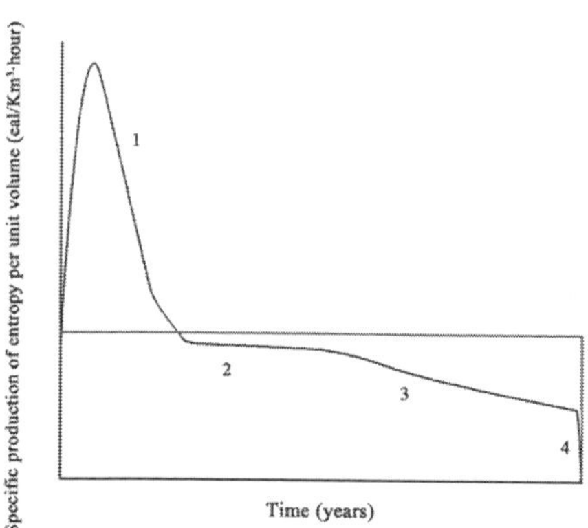

Figure 2. Schematic diagram of specific (per unit volume) production rate of entropy vs. time showing four chronological sta- ges; 1—initial growing, 2—stationary or steady-state stage, (adult- hood), 3—aging in the mature stage, and 4—catastrophic failure or destruction. The rectangle should have the same area to apply Calculus mean value theorem.

It is widely observed that most living dissipative systems perform functional cycles. For example in mammals breathing, heartbeat, digestive processes, etc. are cyclic processes. Of these, the most important functional cycle is the relative to the rate of arrival of nutrients and oxygen to each and every one of the cells that make up every living system. The heartbeats promote, throughout the body, a distribution of free energy per unit time which allows meeting the needs of the dissipative system. Clearly, for each process that occurs within the system, the length of each cycle varies with the age of each living being and that may depend on the specific activities carried out at all times. As a first approximation, the case of the cyclic dissipative systems will be analyzed, under the assumption that a large population of them can be represented by average values through their working life. Based on the second law of thermodynamics, is considered here that any living being gradually wears down and eventually stops working as it previously did (in a similar way that occurs for an internal combustion engine, which also operates through cycles).

During their operation dissipative systems perform internal processes of a cyclic nature that could be repeated hundreds, thousands or thousands of millions of times, etc. The dissipative structures go through a maximum number of cycles, N_{max} before a catastrophic failure occurs. Each type of dissipative system has its own characteristic maximum number of cycles. In other words, let us state this important general fact: Every dissipative structure working by functional cycling necessarily generate a net quantity of functional entropy or functional disorder in every cycle and, therefore, the structure cannot last longer than a maximum time of continuous work characteristic of each structure. In other words, the total quantity of energy dissipated as heat by any dissipative structure during its total working time has a maximum value. This finite time of continuous operation is due to the increase of functional entropy or functional disorder which eventually causes the breaking of the dissipative machine. We will show that any dissipative system realizing cycles obeys a maximum action principle, which value is characteristic of each system.

If we consider the whole process to take place at a constant temperature T , then from Figure 2, it is clear that for each particular dissipative system, the area under its curve is the maximum generated entropy. In the following it will be considered that when a species has reached its evolutionary stability, their adult members have common dissipative thermodynamic characteristics. Then, the statistical description of any of its living members (any age) can be obtained by using estimation or measuring the maximum total free energy dissipated through his life of the individuals who have been the longest-lived of the species.

In mathematical language

$$\int_{t_i}^{t_f} \frac{ds}{dt} dt = \frac{K_s}{T} \quad (7)$$

where K_s is the energy dissipated during the whole working process for each dissipative type of machine or living system, and t_i and t_f are the initial and final times, respectively. If we make use of the Calculus mean value theorem, then

$$\overline{\frac{ds}{dt}}(t_f - t_i) = \frac{K_s}{T} \quad (8)$$

where $\overline{\frac{ds}{dt}}$ is the average value of the production rate of specific entropy for our system between times t_i and t_f, where the graphic meaning of Equation (8) in the scheme of Figure 2 is obvious. By using Equation (5), for average values, into Equation (8); we have that

$$\frac{\overline{\varnothing_R}}{T}(t_f - t_i) = \frac{K_s}{T} \quad (9)$$

Thus, the total operation time of the dissipative system is given by

$$(t_f - t_i) = \frac{K_s}{\overline{\varnothing_R}} \quad (10)$$

Which shows in a direct way that a dissipative system has a maximum of continuous time operation inversely proportional to the specific dissipative Raleigh's function of the system, and directly proportional to their K_s, which is characteristic of each dissipative system.

3.2. Cyclic Nature of Dissipative Systems

It is clear that many dissipative structures realize functional cycles, which repeat over and over during its whole period of operation, before fail. Assuming, for simplicity, that the cycles are identical, then, from Equation (9):

$$\frac{\overline{\varnothing_R}}{T}\tau\left[\frac{t_f - t_i}{\tau}\right] = \frac{K_s}{T} \quad (11)$$

Here τ is the average time which characterizes the cycle duration. Equation (11) means that every time the living machine performs one of the $N_{max} \equiv \left[\frac{t_f - t_i}{\tau}\right]$ cycles, it produces a quantity of specific entropy, $\frac{\overline{\varnothing_R}}{T}\tau$.

In the sake of mathematical simplicity, constant or average duration of the functional cycles through life will be assumed, even though we know that during adulthood metabolism decreases in time, which causes that the time duration of cycles also diminish.

The existence of an upper limit to the number of the operation cycles implies also the existence of accumulative damage and aging processes. This damage and aging processes cause the continuous decreasing on the power that the dissipative can handle through time and from empirical observation is clear that the aging rate is different and characteristic for each type of system.

For auto-organized systems, there are many phenomenological observations, which show that the macroscopic symptoms of aging begin to appear after the auto-organized dissipative structure attains adulthood [13] . This stage corresponds to the stage 2 of Figure 2 previously defined. In the stage 3, the specimen already arrived to adulthood, and the production rate of specific entropy will be a decreasing function of time.

Taking into account such empirical information from macroscopic signs of aging it is possible to make an approximation about the derivative in time $\dot{s}$, which can be expressed as a function relative to the steady-state rate of entropy production $\overline{(\dot{s})}$ like a bilinear function of α and t,

$$\dot{s} = \dot{s}(\alpha, t) \tag{12}$$

where α is an aging parameter measuring the functional decay of the dissipative system in each cycle. It is obvious that the simplest expression capable to deal with the decreasing on the rate of entropy production by aging in stage 3 is

$$\dot{s}(t) = \overline{(\dot{s})} - \alpha * \overline{(\dot{s})}\left[\frac{t - t_{sd}}{\tau}\right] \tag{13}$$

Here t_{sd} indicates the time where steady-state begins to decay and t_{sd} corresponds to the beginning of stage 3. Equation (13) becomes:

$$\dot{s}(t) = \overline{(\dot{s})}\left[1 - \alpha\left(\frac{t - t_{sd}}{\tau}\right)\right] \tag{14}$$

which clearly states that every dissipative structure necessarily suffers accumulative damage during its operation. We can say that the dissipative system has some functional memory, in the sense that cannot be rejuvenated. That is, $\dot{s}(t)$ decreases with the accumulated time of operation of the dissipative system at a rate given by α.

Neglecting the contribution of the collapse stage 4 (assuming that this last stage is extremely short as compared with the stage 3), also from Equation (14) it is easy to define a new physical macroscopically variable related to the steady-state of entropy production; the fractional remaining functionality:

$$\Pi_{re}(t) \equiv \frac{\dot{s}(t)}{\overline{(\dot{s})}} \tag{15}$$

Therefore,

$$\Pi_{re}(t) = 1 - \alpha * \left(\frac{t - t_{sd}}{\tau}\right) \tag{16}$$

From this equation, it is clear that when the dissipative system collapses, $\Pi_{re}(t_f) = 0$. This result implies that α can be evaluated, for instance, from the total failure condition as follows:

$$\alpha = \frac{\tau}{(t_f - t_{sd})} \tag{17}$$

By using Equation (17) into Equation (14)

$$\dot{s}(t) = \overline{(\dot{s})}\left[1 - \left(\frac{t - t_{sd}}{t_f - t_{sd}}\right)\right] \tag{18}$$

which describes the time evolution of $\dot{s}(t)$ after steady-state ceases $(t > t_{sd})$.

Also, it is possible to define a functional damage, $D(t)$, as

$$D(t) \equiv \left(\frac{t - t_{sd}}{t_f - t_{sd}} \right) \tag{19}$$

which could be expressed as a function of the total cycle number at failure condition N_T characterizing the duration of each structure:

$$D(t_n) = \frac{n}{N_T} = \frac{W_n}{W_{N_T}} \tag{20}$$

Here w is the work per cycle, $W_{N_T} \equiv N_T w$, and $W_n \equiv nw$; with N_T as the total cycle number at failure condition. By using Equations (17)-(19) in Equation (16) we get

$$\Pi_{re}(t) + D(t) = 1 \tag{21}$$

This equation shows, as expected, that fractional remaining functionality has high values, when the functional damage is low. For animals, these equations allow to see that at the time t, $(t > t_{sd})$, the specific power capable to be used or assimilated by the dissipative system, $\Pi_{re} * \overline{(\dot{s})T}$, is lower than $\overline{(\dot{s})T}$; because of the thermodynamical limitations imposed by the functional damage $D(t)$ produced by the operation of the dissipative system itself. It is clear, that $D(t) * \overline{(\dot{s})T}$ is the specific power that the dissipative system is incapable to use or assimilate, after steady-state, because of the accumulated functional damage at time .

3.3. Maximum Action Principle

Here, by using our formalism, it is possible to obtain an equation for the maximum entropy changes in a dissipative system as multiples of the Boltzmann's constant. For this purpose, let us employ another physical meaning of Equation (9) if we use Equation (5), together with $\dot{s} = \frac{dQ_i}{dt}$ as the specific entropy production rate for the i-dissipative system, it is possible to obtain the following equation:

$$\dot{\theta}_i t_{fi} = K_{si} \tag{22}$$

where $\dot{\theta}_i = \frac{1}{M}\left(\frac{dQ_i}{dt}\right)$, Q_i is the dissipated heat, M is the mass of the i-system and t_{fi} is the failure time of the i-system. Therefore, Equation (22) can be written as

$$dQ_i t_{fi} = MK_{si}dt \equiv h_{i\max} \tag{23}$$

where $MK_{si}dt \equiv h_{i\max}$ is a constant for each i-dissipative system and its physical meaning is the maximum action developed by the dissipative i-system during their total dissipative operation. Recall that the action concept which has dimensions of [ENERGY multiplied by TIME], comes from mechanics in relation to Integral Variational Principles of Least Action, which give place to the motion equations for any mechanical system [14] .

Therefore, the total entropy generated during the continuous dissipative operation of the dissipative system, ΔS_i, as a function of the maximum action, is

$$\Delta S_i = \frac{1}{T}\left(\frac{h_{i\max}}{dt}\right) \tag{24}$$

Notice that physically and mathematically Equation (23) resembles the uncertainty principle due to Heisenberg;

$$\Delta E \Delta t \geq \frac{h}{2\pi} \tag{25}$$

where E is for the energy and t for time, and h is the Planck's constant.

By using Equation (25), into Equation (24) it is possible to obtain

$$\Delta S_i = k\left(\frac{h_{i\max}}{h/2\pi}\right) \tag{26}$$

where k is the Boltzmann's constant. Therefore, the total entropy generated during the continuous dissipative operation of the dissipative system, ΔS_i, is a function of the ratio between the maximum action characteristic of each type of dissipative system, and the Planck's constant which is the minimum quantum action [14] , times the Boltzmann's constant.

In the following paragraphs, a comparison between the theoretical results here obtained with experimental data, and other theoretical considerations will be made.

4. COMPARISON BETWEEN EXPERIMENTAL DATA AND THEORY

4.1. Average Lifespan of Human Females and Males

First, we will give a quantitative thermodynamic explanation to experimental facts that on average (for all current country populations) women live several years more than men. Later, we will discuss some thermodynamic implications of aging of humans and their internal organs. On one hand, fromFigure 3, [15] it is immediate that the average basal metabolic rate in man throughout his existence is 40 $(\mathrm{Kcal/sq \cdot m/h})$, and for women is 36.8 $(\mathrm{Kcal/sq \cdot m/h})$. According to [16] the mean body area for European man and female are 1.9 m^2 and 1.6 m^2 respectively. Also from reference [17] the average European male weight is 78.9 Kg and for female the value is 65.8. Therefore, the average value for the basal metabolic rate per unit weight (bmr) for European man and women are $0.963\ (\mathrm{Kcal/Kg/h})$ and $0.9161\ (\mathrm{Kcal/Kg/h})$ respectively. Using Equation (10) with the consideration that for this case K_s (which is the energy dissipated during the whole working process for each dissipative type of living system has the same value for the human species), we have the following theoretical expression for the ratio between the span of life of European human females $t_{\mathrm{SL\ Fem}}$ and the span of life of European human males $t_{\mathrm{SL\ Male}}$:

$$\left(\frac{t_{\mathrm{SL\ Fem}}}{t_{\mathrm{SL\ Male}}}\right) = \frac{0.963}{0.9161} = 1.051 \quad ,(27)$$

On the other hand, taking into account the female and male life expectancy of all the members of the European Union (EU) [18] and their respective

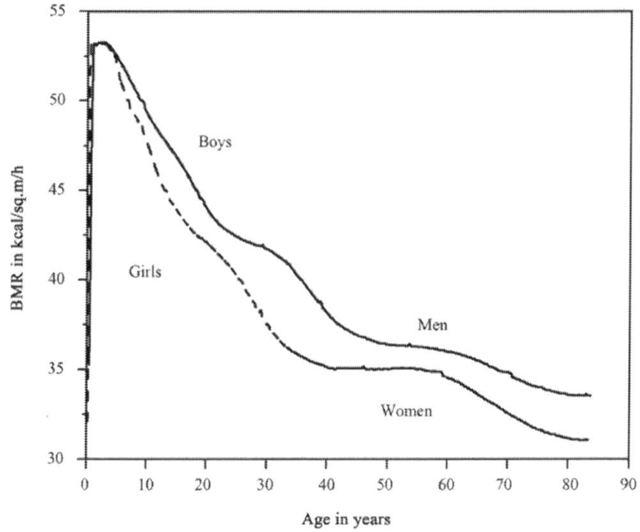

Figure 3. Change in BMR with age [15] .

population, we obtain the average ratio for the female and male life expectancy for the EU,

$$\left(\frac{\overline{\tau_{\text{LE Fem}}}}{\tau_{\text{LE Male}}}\right) = 1.077 \tag{28}$$

Therefore, the percentage difference between the theoretical and the experimental data is 2.5%, which is an excellent result.

There are two reasons for employing EU data; their data availability and very few premature deaths of that relatively rich, healthy and homogeneous population as compared to other regions of the world population. Premature deaths are due to crimes, war, disease, and malnutrition. For instance, in Sub-Saharan Africa, one of the worse health problems (if not the worst) is underweight infancy.

4.2. Aging Rates of Human Organs

Relative to the thermodynamic implications of aging of humans and their internal organs, in Figure 4, experimental data for Average Percent Physiological Function Remaining of different human organs and physiological functions, (as given by Cutler [13]) are shown as a percentage

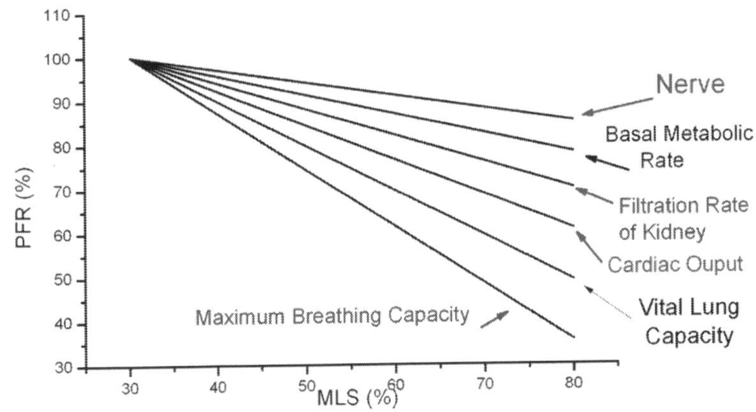

Figure 4. Experimental data for the average percentage remaining functionality (PFR) for different human organs and functions as a function of the percentage of maximum lifespan potential (as given by Cutler [13]). We point out that these straight lines also appeared in the original work.

Table 1. Aging parameters, a, which characterize the functional decay of the dissipative system in each cycle corresponding to different organs in human bodies.

Organ or Human Function	Values of α
Nerve	$\alpha = 9.38 \times 10^{-11}$
Basal Metabolic Rate	$\alpha = 1.388 \times 10^{-10}$
Filtration Rate of Kidney	$\alpha = 1.898 \times 10^{-10}$
Cardiac Output	$\alpha = 2.482 \times 10^{-9}$
Vital Lung Capacity	$\alpha = 3.22 \times 10^{-10}$
Maximum Breathing Capacity	$\alpha = 4.1 \times 10^{-10}$

of maximum lifespan potential; In principle, all these organs obey the Equation (16), but since they form part of the whole system, in reality, they are interdependent.

From data of Cutler et al. [13] as shown in Figure 4, by using Equation (16) we can calculate the values for the aging parameter, α (for each organ). If we use the same failing time and cardiac cycle τ for each organ, then the ratios of different α do not depend on this choice. We choose $\tau = 1\ \mathrm{s}$ and $\text{failing time} = 100\ \text{years}$, the results are shown in Table 1.

The data on Table 1 allow to describe the time dependence of the percentage functional remaining, π_{re}, corresponding to different organs, from aging human bodies, is in full agreement with experimental data. According to the theory of evolution of species in a species that has reached stability, all internal organs (in general) are aging at the same rate. Therefore all deviation from that behavior obeys environmental, behavioral or dietary factors outside the prevailing natural diet and behavior in the environment where the species evolved. One pos- sible factor for the variations on slopes in π_{re} function for different human organs with living time could depend to social conducts like smoking, taking of alcohol, overweight, etc., that causes deleterious effects on their operation. A physical interpretation of such results is as follows: The rate of aging of the nerves is 3.43 times smaller than the average rate of aging social vital lung capacity, or about 2.64 times lower than the social average rate of aging for the cardiac impulse.

5. CONCLUSIONS

We identified general four chronological stages of self-organized living dissipative structures of the specific (per unit volume) production rate of entropy: 1) an initial more or less exponential transient stage, 2) stationary or steady-state stage, (adulthood), with slow aging, 3) aging in the mature stage" with an approximate linear decrease of $\dot{s}$ with time, and 4) catastrophic failure or destruction stage.

Based on the proposal of an average dissipation rate model for non-linear irreversible thermodynamics for dissipative structures, which goes through functional cycles, we conclude that the systems necessarily generate a maximum quantity of entropy (characteristic of each type) and, therefore, last a maximum quantity of time t_{fi}. Such analysis allowed us to develop a new theory to describing the essential aspects of aging processes for any kind of dissipative systems.

The second law of thermodynamics has been used to explain the finite time duration of stationary states and simultaneously, both the processes of intrinsic progressive loss of functionality for non-living dissipative systems, and the processes of intrinsic progressive loss of organic function in the living dissipative systems. Some of the theoretical results (as Equation (10)) show in a direct way that all kinds of dissipative system have a maximum of continuous time operation inversely proportional to the specific dissipative

Raleigh's function of the system, and directly proportional to their K_s, characteristic of each specific type of dissipative system. This equation allows us to show in a direct and quantitative way why on average the human female last longer than the human males.

We hope that future developments of simple concepts developed here, such as fractional remaining functionality (Equation (16)), could to make a contribution to develop programs for preventive health care. Different values for corresponds to distinct aging rates on a given subsystem of human bodies. Then for a given organ (fundamental or not) it is possible to collect data of a human population and select the people with lower values for α than the population average, in order to make studies about their living and eating habits in order to develop preventive measurements for the rest of the population. It is clear that the previous analysis, for average population basal metabolism, in other words is a statistical scheme and it is not possible to distinguish personal behavior as for example: smoking, drinking or homosexual personal conducts.

Figures of the type of Figure 4 and the use of Equation (16), in principle allow to compare the general state of health of a concrete human being of x age with regard to the general state of health at the same age corresponding with an average of humans being that later have arrived to the 100-year-old age. And also to determine the maximum duration of the person's life that moves away from the values that correspond to the possibility of arriving to a 100-year-old age. In short, for example, after evaluating the alpha corresponding to the centennials and knowing the average of the numerical values for the heart functionality in a 100-year-old population we can determine the value that would throw the use of Equation (16) for a specific person of x age. It would allow it to evaluate the difference among the heart functionality from the centennials to that age and the subject in question. If we consider that the value of the heart functionality of the centennials is a limit value; then the use of Equation (16) for a specific person will determine us statistically their maximum duration of life. This analysis type can be a complementary useful tool, cheap and effective that cooperates to the standard evaluation procedures of the future health of big groups of human beings. For the time evolution of the basal metabolism on children is necessary a further independent study.

Analogous to the quantum Heinsenberg's principle, we presented an empirical principle for maximum action of any dissipative system involving the Planck's constant.

We made use of average values of the basal metabolic rate per unit weight of European man and women to conclude that, on average, the human female last longer than the human males, with only a 2.5% of relative error.

Also, we calculated the aging parameter measuring the functional decay of the dissipative system in each cycle from specific production of entropy of some human organs or physiological functions.

We assumed constant duration of the cycles in adulthood, even though we know that during adulthood, metabolism decreases in time. Then it will be desirable to refine our formalism to include different values of the average and the variance of the duration of the cycles in different stages of the living structures.

The substitution of organs as a way to increase the span of human life, we would have as limit the time implied by the aging rhythm of the original brain. In the same way that the inclusion of new pieces in an old car does not increase to infinite the lifespan of the car, since during the replacement process, a stage will be attained were the car rejuvenated no longer corresponds with the original car.

All microscopic theory of the aging of living beings should be consistent with the second law of the thermodynamics. In other words, the operation of the biological self-organized structures, is not more than an efficient mechanism of degradation of the free energy that we receive from the sun, and it only implies a delay in which the dissipative biological systems outside of equilibrium approach inexorably to the thermodynamic equilibrium obeying the second law of the thermodynamics.

ACKNOWLEDGEMENTS

JAMA specially wants to thanks to Prof. M. López de Haro for many years of deep discussions and arguments and by his contribution to final shaping of the ideas here developed.

REFERENCES

1. Schrödinger, E. (1992) What Is Life? With Mind and Matter and Autobiographical Sketches. Cambridge University Press, Cambridge.
2. Prigogine, I. (1945) Etude thermodynamique des phénomènes irréversibles. Thèse d'agrégation présentée en 1945 à l'Université Libre de Bruxelles. Desoer, Liège, 1947. Académie Royale de Belgique. Bulletin de la Classe des Sciences, 31, 600.

3. Prigogine, I. (1969) Structure, Dissipation and Life. Theoretical Physics and Biology, Versailles, 1967. North-Holland Publ. Company, Amsterdam.

4. Glansdorff, P. and Prigogine I. (1971) Structure, Stabilitéet Fluctuations, Masson, Paris. Thermodynamic Theory of Structure Stability and Fluctuations. Wiley and Sons, London.

5. The Nobel Prize in Chemistry (1977) Ilya Prigogine. Award Ceremony Speech. Presentation Speech by Professor Stig Claesson of the Royal Academy of Sciences. (Translation from the Swedish Text)

6. Feinberg, A.A. and Widom, A. (2000) On Thermodynamic Reliability Engineering. IEEE Transactions on Reliability, 49, 136-146.

7. Kirkwood, T.B. and Austad, S.N. (2000) Why Do We Age? Nature, 408, 233-238.

8. Partridge, L. and Gems, D. (2002) Ageing: A Lethal Side-Effect. Nature, 418, 921-921.

9. Prigogine, I. (1967) Thermodynamics of Irreversible Process. Charles C. Thomas Publisher, Springfield.

10. Glaser, R. (1999) Biophysics. Springer-Verlag, New York.

11. Glansdorf, P. and Prigogine, I. (1971) Thermodynamical Theory of Structures, Stability and Fluctuations. Wiley Interscience, New York.

12. Guytton, A.C. (1976) Textbook of Medical Physiology. 5th Edition, W. B. Saunders Company, Philadelphia-London- Toronto, 952.

13. Cutler, R.G. (1978) Evolutionary Biology of Senescence. In: Behnke, J.A., Finch, C.E. and Moment, G.B., Eds., The Biology of Aging, Plenum Press, New York, 311-360.

14. Sommerfeld, A. (1972) "Mechanics" Lectures on Theoretical Physics, Volume I. Paperback Edition, 6th Printing, Academic Press, Inc., Waltham, 181.

15. Durnin, J.V.G.A. (1981) Basal Metabolic Rate in Man. University of Glasgow Glasgow Scotland. Joint FAO/WHO/ UNU Expert Consultation on Energy and Protein Requirements Rome, 5-17 October. Sparreboom, A. and Verweij, J. (2003) Paclitaxel Pharmacokinetics, Threshold Models, and Dosing Strategies. Journal of Clinical Oncology, 21, 2803-2804.

16. (2006) Health and Food. Fieldwork November, December 2005. Special Eurobarometer. European Commission.

17. WHO (2013) WHO Life Expectancy.

5

CHAPTER

Calorimetric Investigations of Non-Viral DNA Transfection Systems

Kaur Tranum[1], Tavakoli Naser[2], Slavcev Roderick[1] and Wettig Shawn[1]

[1] School of Pharmacy, University of Waterloo, Canada

[2] School of Pharmacy and Pharmaceutical Sciences Isfahan University of Medical Sciences, Isfahan, Iran

1. INTRODUCTION

Although the structural and energetic forces involved in DNA condensation have been studied for years, this area has experienced a resurgence of interest in recent years with respect to developing gene therapy protocols to combat a variety of human diseases. Despite an intense effort to study the mechanism(s) of DNA condensation using a variety of techniques such as microscopic, light scattering, fluorescence, and calorimetric techniques the precise details of the energetics of DNA nanoparticle formation and how these complexes assemble is not well understood at present. Isothermal titration calorimetry (ITC) has become an important technique for studying the binding energetics of biological processes, including protein-protein binding, protein-carbohydrate binding, protein-lipid binding, antigen-antibody binding, DNA-protein binding, and the DNA-lipid binding that is at the heart of non-viral transfection vectors. Given the large informational content of thermodynamic data (such as binding constants, stoichiometry, enthalpies of interaction, etc.) it is not surprising that ITC is being used for the elucidation of the binding mechanism of DNA with surfactants or lipids used in non-viral gene delivery vectors. This is described in more detail in this Chapter.

1.1. Principle and Experimental Set Up

One of the fundamental challenges in biophysical chemistry, as well as in all physicochemical events implying a solvent, is to attribute the contribution of different non-covalent interactions (electrostatic in the most general sense, solvation and hydrophocic interactions, hydrogen bonding, Van der Waals interactions) to the free energy change of a given molecule upon its interaction with a binding partner. Another challenge is to investigate how conformational changes in the three dimensional structure of a protein, DNA or RNA can modify the interaction between itself and its binding partner. Finally, the most important challenge is to relate structure (the high resolution X-ray or NMR structure of more and more proteins and nucleotides are available) to the "binding affinity" of these molecules with their binding partners. This means that the "binding affinities" have to be accurately measured for as many as possible systems to get a clear picture of structure-activity relationships, which is particularly important for the design of new drugs or the design of gene vectors for gene therapy. Unlike any other method used to determine thermodynamic parameters associated with bimolecular interactions, calorimetery uses the direct measurement of the heat of interaction (or molar enthalpy ΔH) to probe the extent of binding between the molecules. When substances bind, heat is either generated or absorbed.

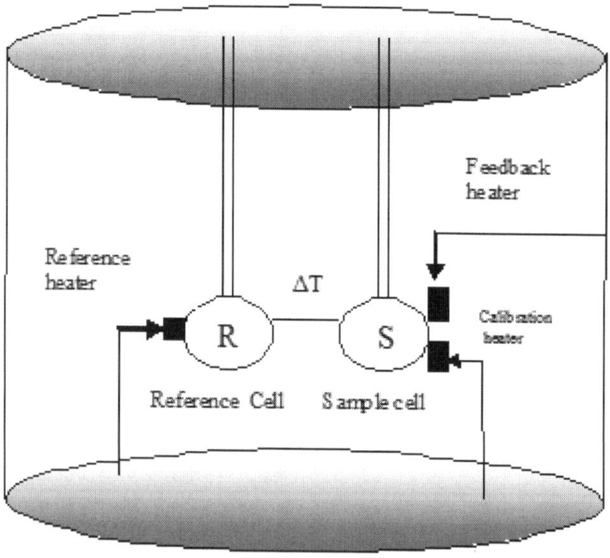

FIGURE 1. Schematic diagram of an ITC instrument.

Isothermal microcalorimtery is a thermodynamic technique that directly measures the heat released or absorbed during a biomolecular binding event. Measurement of this heat allows accurate determination of binding constants, reaction stoichiometry (n), enthalpy (ΔH) and entropy (ΔS), thereby providing a complete thermodynamic profile of the molecular interaction in a single experiment. The first modern isothermal microcalorimeter was designed by Calvet (Calvet and Prat, 1963). An ITC instrument consists of two identical cells composed of a highly efficient thermal conducting material (e.g. gold) surrounded by an adiabatic jacket (Figure 1). The jacket is usually cooled by a circulating water bath. Sensitive thermopile circuits detect temperature differences between cells and the jacket, while heaters located on both cells and the jacket are activated when necessary to maintain identical temperatures between all components. In a typical ITC experiment, the macromolecule is placed into the sample cell of the calorimeter and is titrated at constant temperature with the ligand in a syringe. The reference cell contains water or buffer. Prior to injection of the titrant, a constant power (< 1 mW) is applied to the reference cell. This signal directs the feedback circuit to activate the heater located on the sample cell. This represents the baseline signal. During the injection of titrant into the sample cell, heat is taken up or evolved depending on whether macromolecular association reaction is endothermic or exothermic. For an exothermic reaction, the temperature in the sample cell will increase, and the feedback power will be deactivated to maintain equal temperature between the two cells. For endothermic reaction, the reverse will occur.

The heat released (in terms of molar enthalpy ΔH), upon interaction of the titrant is monitored over time. Each peak represents a heat change associated with the injection of a small volume of ligand solution into the ITC reaction cell. As successive amounts of the ligand are titrated into the ITC cell, the quantity of heat absorbed or released is in direct proportion to the amount of binding. When the system reaches saturation, the heat signal diminishes until only heats of dilution are observed. A binding curve is then obtained from a plot of the heats from each injection against the ratio of ligand and binding partner in the cell. A molecular interaction between two ligands can be defined by the following equation which forms the basis for an ITC analysis:

$$\Delta G = -RT \ln K_A = \Delta H - T\Delta S \quad (1)$$

The first part of this equation implies that the change in the Gibbs free energy (ΔG) correlates with association constant K_A because R, the gas constant, and T, the absolute temperature, are constant. The dissociation constant K_D, which is commonly used to quantify the affinity between two ligands, is the inverse of K_A. The second part of this equation illustrates that the sum of enthalpy (ΔH) and entropy (ΔS) changes define the free energy (ΔG) and consequently the affinity of an interaction. A major advantage of ITC experiments is the fact that K_A and ΔH are measured in a single experiment. Having measured these two parameters the remaining variables ΔG and ΔS can be derived. In combination with structural information, the energetics of binding can provide a complete picture of the interaction and aid in identifying the most important regions of the binding interface and the energetic contribution (Pierce *et al.*, 1999).

2. NON-VIRAL DNA TRANSFECTION SYSTEMS

An emerging trend in the treatment of disease is gene therapy, a broad term that includes any strategy to treat a disease by delivering exogenous gene, gene segments, or oligonucleotides into the cells of patients in order to restore normal functions to the cell. Such therapies are either DNA based (usually in the form of plasmids) or RNA based (usually in the form of small, interfering RNA or siRNA), both requiring some form of delivery system for the oligonucleotide(s) in question.

Two major classifications of delivery systems exist for DNA therapies, viral- and non-viral-based. Viral vectors typically demonstrate the highest gene transfer activity; however, they suffer from serious safety concerns particularly with respect to inflammatory and immune responses in patients. Non-viral vectors are generally comprised of a combination of cationic lipids, neutral lipids and/or cationic polymers, and have a much better safety profile compared to viral systems but suffer from low gene transfer activities. As such a great deal of effort is being expended towards the development of new non-viral vector systems with increased transfection capabilities.

Lipoplexes, complexes of liposomes with DNA, are the most investigated non-viral transfection vectors, and hold great potential application for gene delivery. They have been utilized in approximately 8% of the clinical trials for gene therapy mainly due to their advantages over viral vectors. They show low toxicity, high therapeutic gene capacity, cell-specific targeting capability, and straightforward production, modification, and functionalization procedures. A key problem in the advancement of lipoplex transfection vectors, yet to be efficiently overcome, is the above mentioned inadequacy of transfection efficiency of current cationic lipid systems, attributed to the incomplete understanding of interaction mechanisms involved. While this is driving research efforts toward the design and synthesis of novel amphiphilic cationic compounds, as well as approaches for formulation of liposomal based vectors (Giatrellis *et al.*, 2009), less effort is being directed towards better understanding the fundamental mechanism(s) by which liposome-mediated transfection occurs.

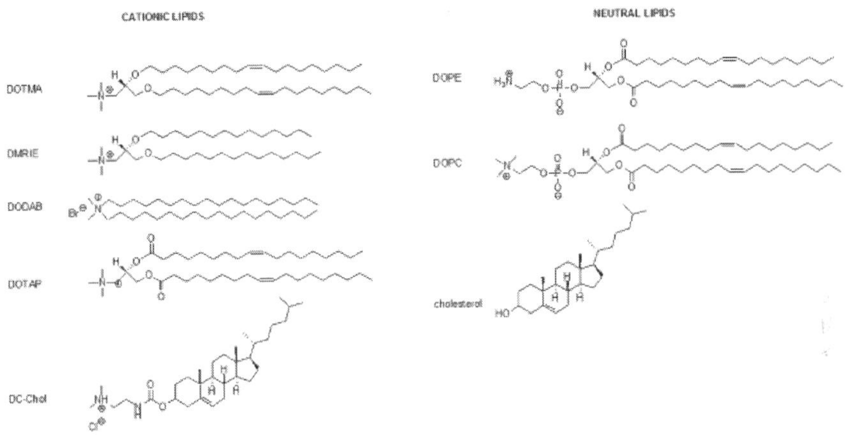

Scheme 1. Chemical structure of lipids commonly used in the formulation of DNA transfection complexes: DOTMA (1,2-di-O-octadecenyl-3-trimethylammonium propane), DMRIE (1,2-dimyristyloxypropyl-3-dimethyl-hydroxy ethyl ammonium bromide), DODAB (dioleyldimethylammonium bromide), DOTAP (1,2-dioleoyl-3-trimethylammonium-propane (chloride salt), DC-Chol (3β-[N-(N',N'-dimethylaminoethane)-carbamoyl] cholesterol hydrochloride, DOPE (1,2-dioleoyl-sn-glycero-3-phosphoethanolamine), DOPC (1,2-dioleoyl-sn-glycero-3-phosphocholine), and cholesterol.

The first reported application of cationic liposomes (or vesicles) as DNA transfection vectors was in 1987, by Felgner et al. They synthesized a cationic lipid, N-[1-(2,3-dioleyloxy)propylJ-N,N,N-trimethylammonium chloride (DOTMA), that forms small unilamellar cationic liposomes, closed bilayer membrane shells of lipid molecules, and that interact with DNA spontaneously via electrostatic attraction. These lipid-DNA complexes maintain an overall positive charge, facilitating uptake by cells again by electrostatic attraction, now between the cationic complexes and the predominantly negatively charged cellular membrane. Complexes are then internalized allowing the transfer of functional DNA into the cell followed by expression. The technique was simple, highly reproducible, and more efficient than some other, commonly used procedures such as use of polycations, calcium phosphate, microinjection, electroporation and protoplast fusion (Felgner *et al.*, 1987).

2.1. Mechanism of Dna – Lipid Interactions

Due to electrostatic interactions between negatively charged phosphate groups along the DNA back bone and the cationic head group of the lipid, they compact the DNA plasmid into small particles. This compaction serves two purposes:

1) to more easily transport the DNA across negatively charged cellular membrane(s), both in terms of eliminating electrostatic repulsion between negatively charged DNA and the cell membrane (Barreleiro et al, 2000, Safinya, 2001), and also by simply having a smaller size (relative to an uncomplexed DNA plasmid); and

2) to protect the DNA from degradation (primarily by DNAses) during the transfection process (Tranchant *et al.*, 2004;Elouahabi and Ruysschaert, 2005).

The importance of the cationic lipid is however not restricted solely to the charge of its head group. (Cherezov *et al.*, 2002). Kreiss et al. believed that both the nature (or structure) of the cationic lipid as well as the DNA backbone exert a strong influence on lipid-DNA packing and complex morphology, with the formation of the lipoplexes resulting from competitive interactions between electrostatic forces of lipid-DNA and elasticity forces driven by the lipid hydrophobic moiety (Zuzzi *et al.*, 2007;Ma *et al.*, 2007;Kreiss *et al.*, 1999).

It has been suggested that lipoplex formation is an endothermic process and the repulsion force between DNA molecules as well as the release of tightly bound counterions from DNA and the lipid mainly controls the formation of cationic lipid/DNA complex (Ma *et al.*, 2007). In the course of complex formation, DNA acts as a polyanion molecule interacting with the cationic headgroups at the surface of liposomes. The neutralization of charges seems to affect both the DNA and the liposome structures. First, DNA might adopt a more compact structure, which is less accessible to intercalating dyes or nucleases. Second, neutralization might induce lipid bilayer mixing and/or aggregation resulting in fusion of liposomes and/or multilamellar structure formation. The fact that different structures are observed in the same lipoplex preparation further complicates the interpretation (Zabner*et al.*, 1995). Apart from the electrostatic interactions between positively charged lipid head groups and the negatively charged

DNA phosphate backbone, variations in lipid-lipid and DNA-DNA ionic repulsive forces, attractive lipid-lipid hydrophobic interactions, hydration forces and other structural properties of the liposome and plasmid DNA can result in a wide variety of macromolecular structures that can vary not only with concentration(s) but also as a function of time. Hence, upon mixing of cationic lipid and DNA, the supramolecular organization of the two components changes considerably. Despite these changes in organization, the molecular structures of both components are generally preserved, leading to a so-called multilamellar complex, in which DNA molecules are intercalated between intact lipid bilayers and form a tightly packed grid (Smisterova *et al.*, 2001;Pitard *et al.*, 1999;Cherezov *et al.*, 2002;Radler *et al.*, 1997).

2.2. Dna / Lipid Complex Morphologies

The most commonly complex assemblies include DNA molecules sandwiched within liposomal bilayers, DNA electrostatically adsorbed onto the vesicles outer surface, DNA encapsulated in the aqueous phase of the liposomes, and DNA coated by a monolayer of cationic lipid envelop (Felgner and Ringold, 1989;Giatrellis *et al.*, 2009). Therefore, lipoplexes have a broad range of morphologies from a hypothesized "beads on a string" arrangement proposed initially by Felgner (Felgner and Ringold, 1989) in their seminal paper, to more complicated such as spaghetti and meatballs structure (Sternberg *et al.*, 1994), map-pin structures (characterized by spheroidal heads and tapering pins)(Sternberg *et al.*, 1998)(See Figure 2), multilamellar or inverted hexagonal phase structures confirmed by high-resolution synchrotron small-angle X-ray scattering (SAXS) experiments (Safinya, 2001) and sliding columnar phase (O'Hern, 1998).

These above mentioned morphologies; however, are less common than the lamellar, "sandwich" structure (Koltover *et al.*, 1998;Sternberg *et al.*, 1994). Under lipid excess conditions there is only macroscopic heterogeneity (i.e., lipoplexes may be of different sizes) whereas the microscopic structure of lamellar cationic lipid-DNA "sandwich" is uniform as demonstrated by diffraction techniques (Radler *et al.*, 1997).

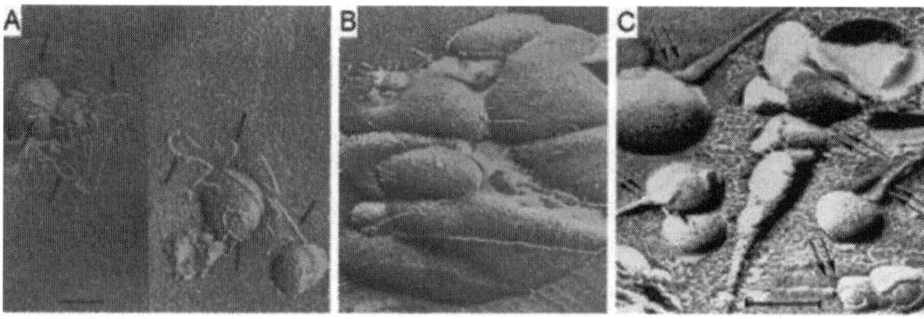

FIGURE 2. A and B showing spaghetti-meatball assembly [adapted from (Sternberg *et al.*, 1994)] and C showing map-pin structures [adapted from (Sternberg *et al.*, 1998)].

2.3. Physiochemical Properties, Lipoplex Structure, and Their Impact on Transfection Activity

Structure-activity studies suggested that biophysical properties, such as size, charge, and morphology of the resulting DNA complexes determine transfection efficiency within one class of vector. Several parameters effect transfection activity, such as structural variations of the cationic lipids (length and degree of unsaturation of the alkyl chains), the nature of the groups bound to the quaternary nitrogen, and the counterions. Variations in the lipid to DNA charge ratio and the presence or absence of helper lipids are other important formulation parameters of transfection efficiency (Lobo *et al.*, 2002;Maloneet *al.*, 1989). In the majority of reported studies, lipid-DNA complexes function most effectively when a cationic lipid is mixed with a neutral or helper lipid such as DOPE (1,2-dioleoyl-sn-glycero-3-phosphoethanolamine) or DOPC (1,2-dioleoyl-sn-glycero-3-phosphocholine) (Scheme 1). DOPE is by far the most commonly used helper lipid in non-viral DNA transfection applications (Ewert *et al.*, 2008;Hui *et al.*, 1996). The inclusion of neutral or zwitterionic lipids (co-lipids), such as cholesterol, DOPE and DOPC as well as PEG lipids, other cationic lipids, and non-lipid compounds like polymers or peptides in the cationic vesicle formulations has been shown to increase the transfection efficiency of the assemblies (at least in case of DOPE) by promoting membrane fusion in different cell lines (Koynova *et al.*, 2007).

2.3.1. Size and zeta potential

In general, transfection efficiency

i. requires a positive or neutral zeta potential,

ii. is size dependent, e.g., it is higher for smaller size particles, and

iii. requires a vector that is stable in serum.

The size of the DNA plays an important role in biodistribution, cellular internalization and intracellular trafficking (Pathak *et al.*, 2009). The correlation between lipoplex size and DNA transfection has been well studied; however, the most suitable size range (for transfection) still remains to be controversial. Early studies suggested the lipoplex sizes in the range of 400–1400 nm for more efficient transfection in cell culture than smaller (<400 nm) or larger (>1400 nm) aggregates (Kawaura*et al.*, 1998), but later results emphasize a smaller (<200 nm) lipoplexe size range (Zhang *et al.*, 2003). This controversy has been attributed to different size-dependent endocytic pathways involved in uptake of lipoplexes. As the gene transfection by the cationic liposomes was shown to be inhibited by wortmannin, an inhibitor of endocytosis, it is suggested that the vesicles with moderate diameters are useful for gene transfection by endocytosis (Kawaura et al. 1998). Moreover, addition of non-lipid condensing agents such as polyethyleneimine, polylysine, protamine sulfate to the cationic lipid- DNA mixture have been reported to produce small homogeneous lipoplexes, in the range of 100 nm, more suitable to enter narrow capillaries than lipid-DNA alone. Further, the lipoplex particle size also plays a role in DNA release in which for lipoplexes with mean size of ~400–500 nm and lamellar distances of ~5–6 nm (and hence are composed of several tens of layers) extensive intermembrane interactions are required for their disassembly (Koynova, 2009). Zeta potential of lipoplexes is usually measured at different DNA/lipid ratios whereas pure cationic vesicles show a positive zeta potential of ~ 60 mV, and it decreases with addition of DNA and changes to negative at DNA/lipid charge ratio of more than 1 (Koynova *et al.*, 2007). Takeuchi et al showed that transfection efficiency of plasmid DNA (pSV2CAT) into mammalian cultured cells was consistent with the values of zeta potential of cationic liposomes containing a tertiary amino head group(Takeuchi *et al.*, 1996). Therefore, lipoplexes with a small excess of positive zeta charges favours higher transfection efficiency (Elouahabi and Ruysschaert, 2005) and *vice versa*, for lipoplex formulations containing excess amount of DNA and negatively surface charge, the release of DNA is significantly hindered compared to lipoplexes with positive zeta potential (Koynova *et al.*, 2007).

2.3.2. Structure

The role of the cationic lipid molecular structure on transfection activity of resultant lipoplexes has recently been reviewed; with the conclusion that lipids having either saturated alkyl chains with a length of ~14 carbon atoms OR longer, monounsaturated chains having trans-isomerism are generally the most effective for transfection (Koynova and Tenchov, 2009; Koynova *et al.*, 2008). It is important to note that this is only a generalization, and factors such as the stoichiometry of the cationic lipid to DNA, the ionic strength of the formulation, the temperature of the formulation (both during and after formulation), and the incubation time can all have a significant effect on the resulting structure of the lipoplex (Giatrellis *et al.*, 2009).

Association of DNA with cationic lipids in a micellar or liposomal form leads to lamellar organization with DNA molecules sandwiched between lipid bilayers. Although the lamellar phase is the common described structure, as evidenced by small-angle X-ray scattering and electron microscopy, some cationic lipid combined with a hexagonal forming lipid could also result with DNA in an inverted hexagonal structure(Barteau *et al.*, 2008). Despite all the advances in biological and biophysical characterization of cationic lipid-DNA complexes, the relationship of their structural properties to their biological activity is still not well understood. Some earlier studies, especially those carried out on lipoplexes containing DOPE, have proposed the inverted hexagonal structure (H^{C}_{II}) is more efficient in transfection than the lamellar (L^{C}_{α}) phase (See Figure 3) (Koltover *et al.*, 1998;Smisterova *et al.*, 2001). However, more recent studies have shown a general lack of correlation between lipoplex structures and transfection activity. Indeed, high lipofection not only depends on the structure and morphology of the vector assemblies, but also on other factors, (one important example of which is the type of cell being transfected; i.e. primary versus clonal, cancerous versus non-cancerous, etc.). The scope of accounting for all such factors makes it difficult for investigators to put together a comprehensive picture of effects such as variation of both cell types and cationic lipids within one study (Ma *et al.*, 2007).

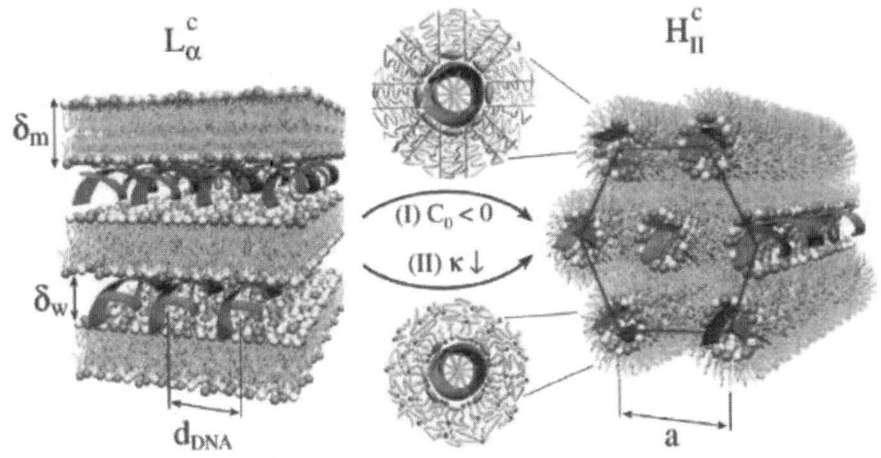

FIGURE 3. Inverted hexagonal structure (H^C_{II}) phase structure and multilamellar (L^C_α) phase structures [adapted from (Koltover *et al.*, 1998)].

2.4. Thermodynamics of Lipoplex Formation

Self-assembly of surfactants into aggregates, such as micelles, vesicles, etc., can be induced by increasing the surfactant concentration to above the critical micelle concentration (CMC) and/or by adjusting the temperature to exceed the critical micelle temperature (CMT). In a typical calorimetry experiment, heats of dilution are recorded. Initially, the dilution of the concentrated surfactant solution (usually well above the CMC) will result in the dis-assembly of micellar aggregates into monomers, and the observed heat of dilution will contain both the heat of solution of the monomer surfactant, and the heat associated with micelle dis-assembling, which is the negative of the heat of micelle formation. As the experiment proceeds, the surfactant concentration in the measurement cell increases to above the CMC, as described in section 1, integration of the measured heats for each injection provides an observed enthalpy of dilution, from which, for a typical surfactant, the enthalpy of micellization (or aggregation), ΔH_{mic} can be obtained from the difference in the pre- and post-micellar enthalpies of dilution. The magnitude and sign (i.e., endo- or exo-thermic) of ΔH_{mic} is governed by the types and magnitude of various contributions (hydrophobic and electrostatic) arising from surfactant-surfactant interactions.

The interaction of cationic aggregates (micelles and/or lipoplexes) with DNA, while being more complex that the aggregation of the surfactants or lipids alone, is governed again by a combination of hydrophobic and electrostatic contributions. The thermodynamics of cationic lipid – DNA lipoplex formation are found to be in most cases, endothermic (Pector *et al.*, 2000;Pozharski and MacDonald, 2002). Since, for any process to be thermodynamically favoured the free energy change must be negative, an endothermic lipoplex formation must be driven by a corresponding increase in entropy (recall equation 1). The interesting question then becomes "what exactly is the source of this entropy gain upon lipoplex formation"? It is widely accepted that the major contribution to this increase in entropy arises from counterion release. Bruinsma first pointed out that lipoplex formation is not simply driven by electrostatic attraction between oppositely charged DNA polymer and lipid membrane, but that in fact the situation is more complex given that both components exist in solution with oppositely charged counterions bound to them that are released upon complex formation (Bruinsma, 1998). Release of these counterions upon complex formation results in an entropy gain that is expected to be, according to Bruinsma, $\sim 1\ kT$ per counterion. Both cationic lipid and DNA become partially dehydrated upon complex formation (Hirsch-Lerner and Barenholz, 1999) resulting in an additional entropy gain due to the additional degrees of freedom acquired by the released water. Hence, the entropy based process is driven through the release of water molecules and of bound counter-ions via the ion-exchange process. It is important to note that the released counterions interact more strongly with water and restrict somewhat the mobility of the water molecules around them, an effect that must lead to some entropy loss. Additionally, the DNA and to a smaller extent the complexing lipids also experience some reduction in entropy because of loss of degrees of freedom due to increased rigidity of the complex relative to the separate components. Nevertheless, combined together the net result is entropically driven lipoplex formulation in general; however additional studies to further elucidate various contributions are still required.

As an example, Pozharski and MacDonald examined the strength of binding of DNA to the cationic lipid matrix upon lipoplex formation (Pozharski and MacDonald, 2003). In this study, the binding free energy was determined by monitoring lipoplex dissociation under conditions of increasing salt concentration. By using relatively short oligonucleotides, the investigators were able to determine the binding energy per nucleotide. A combination of calorimetry and fluorescence resonance energy transfer (FRET) was used in their work. For FRET studies a rhodamine labeled lipid was brought in proximity with carboxyfluorescein (CF) labeled DNA so that CF emission becomes quenched upon lipoplex formation. The decrease in the signal verified complex formation. A Microcal calorimeter was used to determine the cationic lipid-DNA binding enthalpy for the titration of 0.46 mM Dickerson dodecamer (5′-CGCGAATTCGCG) into 25 μM EDOPC (O-ethyldioleoylphosphocholine, cationic lipid) in HE-S buffer (of 20 mM HEPES, 0.1 mM EDTA, and 150 mM NaCl at pH 7.5). Figure 4shows the degree of dissociation of complexes, α, plotted against NaCl concentration, and the heat absorbed at various ratios of DNA/ Lipid (D/L).

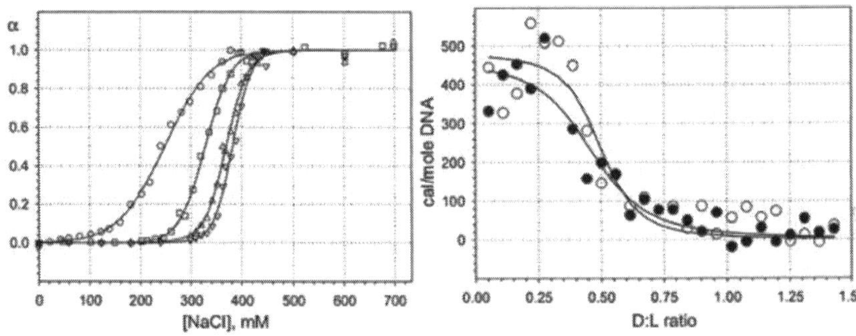

FIGURE 4. DNA-EDOPC binding curves (LEFT) for lipid:DNA charge ratios of: ○2:1 ; □ 4:1 ; Δ 8:1 ; and ▼16:1. Calorimetric profile (RIGHT) of cationic lipid-DNA binding. 460 μM of Dickerson dodecamer titrated into cationic lipid, EDOPC, suspension in HE-S buffer. Curves are from fitting to a one-site binding model; unfilled and solid circles correspond to two independent titrations. [Adapted from (Pozharski and MacDonald, 2003)].

Averaged over two runs, thermodynamic parameters of the binding were $\Delta G = -445 \pm 20$ cal/mole, and $\Delta H = 477 \pm 42$ cal/mole; it was observed that binding becomes more favorable with decreasing cationic lipid:DNA charge ratio. During lipoplex formation there is decrease of the apparent free energy with increasing length of the DNA fragment. It is suggested that there is strengthening of binding at higher lipid:DNA ratios, especially from 8:1 to 16:1, because of the continuing decrease of ΔG up to a charge ratio of 16:1. Lipoplexes with excess lipid actually have stronger lipid-DNA interactions. For example (assuming that the free bilayer is actually part of the lipoplex particle and not as separate vesicles), there may be internal edge effects such that it is easier to accommodate the boundary of DNA-covered areas when there is more lipid.

2.5. Multipoint Nature of Cationic Lipid-Dna Interactions

A DNA molecule binds to the lipid membrane at many points and the relatively small free energies of the individual charges sum to a very large binding energy for the whole molecule. The binding of an individual DNA electrostatic charge to the lipid membrane is rather weak (Pozharski and MacDonald, 2003). Hence, the size of DNA does indeed affect binding energy of lipoplex formation. Single lipid molecules cannot dissociate from the complex since they are held in bilayers by strong hydrophobic interactions and thus form a continuous matrix to which DNA can bind. DNA bases are covalently linked to each other and the stiffness of the polymer is such that a molecule only about as short as a dodecamer binds as a single unit. As a result, much more free energy is released upon a binding event and this brings the apparent dissociation constant into the micromolar range at physiological ionic strength. Longer DNA molecules bind more tightly, but their energy of interaction is also limited by their flexibility. The entropy gain upon lipoplex formation is ~1 kT per released counterion, and in general, this positive ΔS is the driving force of the lipoplex process, which is in good agreement with theoretical considerations. Moreover, very long DNA molecules will exhibit a lower tendency to remain attached throughout their length, but because the number of contact sites is high and because of their increased ability to bend, the extent of dissociation for long DNA molecules at moderate and low ionic strength is extremely small.

3. DNA – GEMINI SURFACTANT SYSTEMS

3.1. Unsubstituted Gemini Surfactants

We have recently reported the results of an isothermal titration calorimetric (ITC) study of the interaction of DNA with transfection systems based upon a family of dimeric surfactants, more commonly known as gemini surfactants (Wettig *et al.*, 2010;Yang *et al.*, 2010;Donkuru *et al.*, 2010). Gemini surfactants consist of two traditional surface active groups (i.e two hydrophobic chains and two polar headgroups) linked chemically by a spacer group. The most extensively synthesized and studied family of gemini surfactants are the N,N-bis-α,ω-alkane-diammonium dibromides, having a general structure [C_mH_{2m+1} $(CH_3)_2N^+(CH_2)_sN^+(CH_3)_2$ $C_mH_{2m+1}\bullet 2Br^-$], hereafter referred as m-s-m, where m represents the carbon chain length of the alkyl tails and s the number of carbon atoms in the polymethylene spacer (See Scheme 2A). Compared to conventional surfactants having a single chain and a single head, gemini surfactants posses unique physiochemical properties, including lower critical micelle concentrations (CMCs) and better wetting properties. However, one of the most important advantages, in so far as gene therapy is concerned, is that the structure of gemini surfactants allows for expanded structural diversity through adjustment of the length of hydrophobic chains, the polarity of head groups, the structure of spacer and the counterions. The spacer length plays perhaps one of the most important roles in determining the properties of these surfactants, and therefore likely their interaction(s) with DNA. The effects of variations in the length of a polymethylene spacer group, as well as the effects of incorporation of hydrophilic substitutents have been extensively studied and are the focus of several reviews on gemini surfactants.

It is generally observed that the chemical arrangement of gemini surfactants provide a rather rich array of aggregate morphologies and solution properties that are dependent upon the nature and size of the linking group. Our previous studies indicated that the transfection efficiency and *in vivo* cutaneous absorption was dependent on the length of the spacer between the two positively charged head groups (Foldvari et al. 2006). Gemini surfactants having a trimethylene spacer demonstrated the highest activity, with the transfection efficiencies correlating to other physical properties (such as the head group area at the air/water interface, critical micelle concentration, etc.) that depend upon the size and/or nature of the spacer group. Moreover, variation in the length of the spacer group allows for a wider variation in the distance(s) between cationic head groups, which provides flexibility for designing surfactants with an optimal match with the anionic phosphate groups on the DNA backbone thereby increasing complexation efficiency.

Microcalorimetry studies used alone and/or along with other techniques can be very helpful in understanding the DNA-gemini interactions in the systems described above. Lipoplexes of gemini compounds, where m = 12, s = 3, 12 and m = 18: 1(Smisterova *et al.*, 2001), s = 2, 3, 6, with DNA were investigated using isothermal titration calorimetry, dynamic light scattering (DLS), zeta potential, atomic force microscopy (AFM) and circular dichroism (CD) techniques. ITC data showed that the interaction between DNA and gemini surfactants is endothermic and the observed enthalpy *vs.* charge ratio profile depends upon the titration sequence(Wang *et al.*, 2007a). Isoelectric points (IP) of lipoplex formation were estimated from the zeta potential measurements and show good agreement with the reaction endpoints (RP) obtained from ITC. DLS data indicated that DNA is condensed in the lipoplex whereas AFM images suggested that the lipoplex morphology changes from isolated globular-like aggregated particles to larger-size aggregates with great diversity in morphology. This work has recently been extended to the study of a non-symmetric gemini surfactant in which one of the alkyl tails is a dodecyl group (i.e. m = 12) and the other has been replaced with a (pyren-6-yl)-hexyl group which has the same approximate length as a dodecyl group, but is significantly more hydrophobic in character (Wang *et al.*, 2007b). The interaction between these pyrene-based gemini surfactants (Scheme 2B) and DNA using ITC (Donkuru *et al.*, 2010), along with the 12-3-12 and 12-6-12 surfactants (for comparison), and marked differences in the interactions with DNA were observed (Figure 5). The 2010 results for the interaction of the 12-3-12 and 12-6-12 with DNA showed a more complex interaction than that originally proposed in 2007, likely as a result of the difference in calorimeters used in the different studies. The major difference observed in the binding interaction(s) with DNA for the pyrenyl gemini surfactants lies in the ability of the pyrenyl group to intercalated between DNA base pairs; this intercalation was previously hypothesized based upon the results of fluorescence titration studies (Wang et al. 2007b).

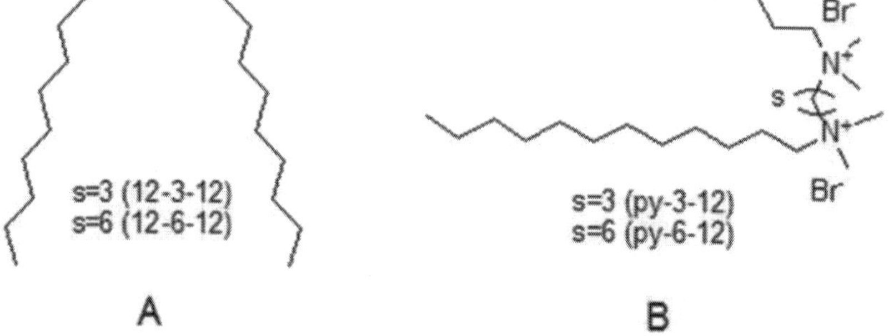

Scheme 2. Structure of the A) 12-s-12 and B) py-3-12 (see discussion below) surfactants.

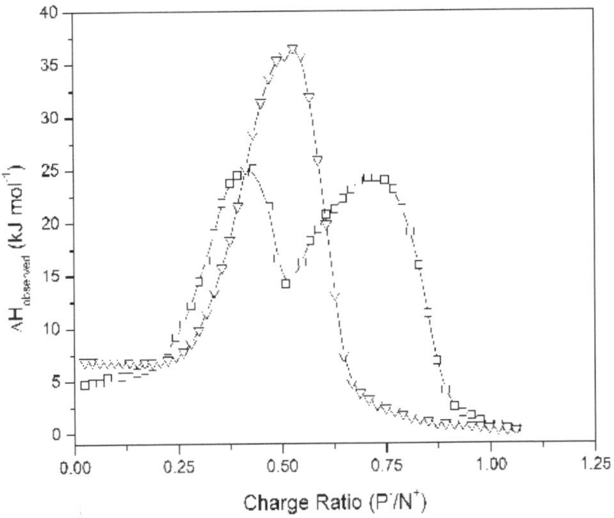

FIGURE 5. Observed enthalpies for the titration of gemini surfactants with DNA: 12-3-12 at 25 °C () and py-3-12 at 25 °C (). Enthalpies are reported in kJ/mol of base pairs. [adapted from (Wettig et al., 2010)]

The double-peak feature observed in the enthalpogram for the titration of the 12-3-12 surfactant with DNA (Figure 5) is common to all of the symmetric gemini surfactants investigated (so far) by our group. The observed binding interactions have been rationalized in terms of the following steps (Wettig *et al.*, 2010):

1. An initial interaction between DNA and surfactant micelles; i.e., formation of the "beads on a string" complex – endothermic;
2. A significant endothermic contribution resulting from the disruption of the micelle-monomer equilibium by continued addition of DNA;
3. A reorganization of the "beads on a string" complex into an approximately neutral complex which gives rise to an exothermic contribution from both the release of structured water and binding between DNA and surfactant monomer;
4. Flocculation of the now neutral surfactant – DNA complexes – endothermic;

5. Precipitation of the complexes (exothermic) followed by no net interaction.

This interaction mechanism is depicted in Scheme 3. The major difference between this mechanism and in the original work published (Scheme 2 in Wang et al., 2007) is that the micelle-monomer equilibrium can no longer be discounted; specifically, the continued addition of DNA to the system has a substantial effect on this equilibrium and results in a significant endothermic contribution to the observed enthalpy. Interestingly a simplification of the system is observed for the pyrenyl surfactants (Scheme 2B), such that a number of intermediate equilibria described above do not seem to occur specifically as a result of the ability for the pyrenyl groups to intercalate between the DNA base pairs.

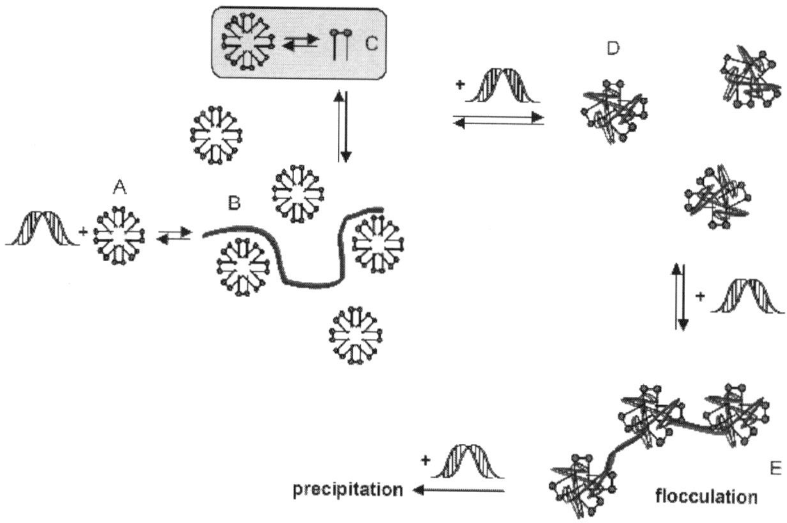

Scheme 3. Proposed interaction mechanism for m-s-m gemini surfactant – DNA systems. Initial binding of DNA to surfactant micelles (A) and formation of the beads-on a string complex (B). Continued addition of DNA results in a disruption of the monomer-micelle equilibrium (C) and the reorganization of the beads-on a string complexes into discrete DNA-surfactant aggregates (D) and eventual flocculation (E) and precipitation of these aggregates. [adapted from (Wettig et al., 2010)]

Other studies have shown that the nature of the counterion significantly affects both micellization and aggregation critical for enhanced trasnfection. The micellization of six cationic gemini surfactants with various counterions, $[C_{12}H_{25}(CH_3)_2N(CH_2)_6N(CH_3)_2C_{12}H_{25}]$ X_2, designated as $C_{12}C_6C_{12}X_2$, with X = F^-, Cl^-, Br^-, Ac^-, NO_3^-, and ½ SO_4^{2-} in aqueous solutions was investigated by microcaloritmetry and conductivity measurements (Jiang *et al.*, 2004). The interaction of these surfactants with DNA in aqueous solutions was also investigated by microcalorimetry. The CMC, CAC (critical aggregation concentration) and the degree of micellar ionization, the saturation concentration of the aggregation, and the associated thermodynamic parameters were detected. Among the counterions examined, $SO_4^{2-}$was the most effective anion for decreasing the CMC (Koynova *et al.*, 2007). Both aggregation processes were mainly entropy-driven since the values of the entropy changes multiplied by temperature were much larger than the absolute values of the enthalpy changes. The binding of micelles to DNA was strongly dominated by the positive entropy gain on release of the small counterions from the micelles and from DNA. Further, the interaction of all of the surfactants with DNA was dependent on the DNA concentration and may be associated with each micelle interacting with more than one DNA molecule. In a another microcalorimetric study, the interaction of a series of dissymmetric gemini surfactants, (designated as $C_{(m)}C_{(6)}C_{(n)}Br_2$, with constant m+n=24, and m=12, 14, 16, and 18) with DNA in 10 mM NaCl was studied and it was found that the dissymmetry degree has a marked effect on the interaction of the $C_{(m)}C_{(6)}C_{(n)}Br_2$ surfactants with DNA since the CAC and saturation concentration tend to become smaller with increased m/n (Jiang *et al.*, 2005).

3.2. Substituted Gemini Surfactant Systems

A more recent focus of research on gemini surfactants deals with the effect of substitution within the spacer group on the aggregation properties of the surfactant, and the rational design of surfactants with improved transfection activity (Kirby *et al.*, 2003;Wang *et al.*, 2004; Wettig *et al.*, 2002; Wettig et al., 2003; Wettig *et al.*, 2007). Transfection activity is found to depend critically upon the structural elements present (Castro *et al.*, 2004). Recently, pseudoglyceryl gemini lipids bearing an oligooxyethylene $(-CH_2-CH_2-O)_m-CH_2-CH_2-)$ spacer were found to be superior gene transfecting agents as compared to those bearing polymethylene $(-CH_2)_m-)$ spacers (Bajaj *et al.*, 2008). In this study, gemini lipids were found to be highly superior in gene transfer abilities as compared to their monomeric lipid and a related commercially available formulation (Bajaj *et al.*, 2008). The increased hydrophilicity of the ethoxylated spacer groups results in an increased water solubility of these compounds, and they preferentially locate at either air-water or micelle-water interface. In some cases, the efficiency of transfection has been correlated to a high extent of DNA condensation, which in turn depends heavily on the gemini spacer structure (Bombelli *et al.*, 2005). An important factor is the ability of the complexes to form polymorphic structures, which are not necessarily hexagonal. Absent from the above, is a thermodynamic characterization of the binding interaction(s) between DNA and gemini surfactants. While the determination of binding properties is complicated by the cooperative nature of the interaction between DNA and gemini surfactants, isothermal titration calorimetry has been used in a few reports.

In order to enhance gene transfection based upon structure-activity relationship, various gemini surfactants (Bombelli *et al.*, 2005) have been designed, synthesized and tested for gene delivery in our laboratory. Previously, we reported the results of a comprehensive study of the aggregation and thermodynamic properties of the 12-4-12 gemini surfactant with and without hydroxyl substitution as well as a series of ethoxylated gemini surfactants. Further, our group is using ITC to extend our work to better understand the thermodynamic properties of the complexation of DNA by the aza-, hydroxyl- and ethoxyl-gemini surfactants, specifically to examine how the binding interactions change as a result of substitution within the spacer group (Scheme 4).

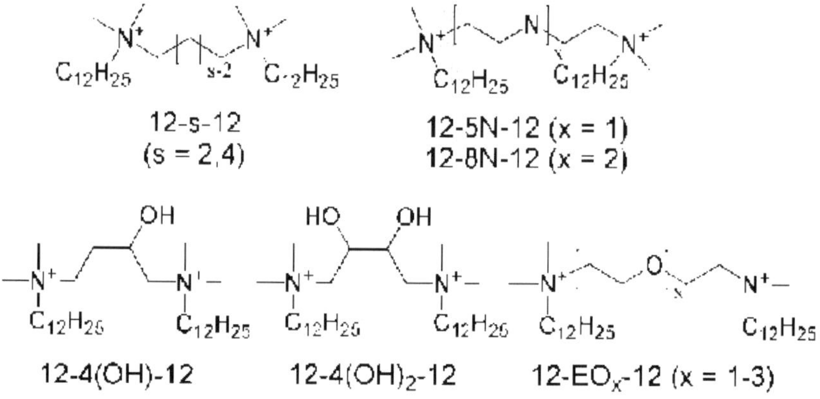

Scheme 4. Structures of the unsubstituted and substituted gemini surfactants.

Enthalpograms for the titration of 2.5 mM bp DNA into 0.5 mM gemini surfactant solutions are shown in Figure 6. The interactions between unmodified gemini surfactants (12-3-12, 12-12-12, 18:1-2-18:1, 18:1-3-18:1, 18:1-6-18:1) (Wang *et al.*, 2007a) and pyrenyl-modified gemini surfactants (Wettig *et al.*, 2010) with DNA have been studied previously in this manner, which provides important information on the nature of the interaction(s) between surfactant micelles and DNA for conditions of excess positive charge; i.e., conditions similar to those typically used to prepare complexes for transfection studies.

The same types of interactions observed between unsubstituted gemini surfactants and DNA (described above) will occur for the various substituted gemini surfactants studied in this investigation; however, since the various substitutions are made within the head group region of the surfactant, this should result in noticeable changes in the binding interactions(s) which is indeed the case as seen in Figure 6. The maximum in $\Delta H_M{}^o$ for the 12-s-12 series has been attributed to energies associated with the configuration of the spacer group at the micelle–water interface. Enthalpies for the substituted gemini's in the presence of DNA were observed to decrease (becomes more exothermic), with increasing hydrophilicity of the spacer group in the order OH < EO < N (Table 1). The increased hydrophilicity of the ethoxylated spacer groups results in an increased water solubility of these compounds, and they preferentially locate at either air-water or micelle-water interface. Very recently, Anissa Bendjeriou-Sedjerar et al. (2009), from their surface tension measurements of gemini surfactants containing two quaternary ammonium groups bound by an ethylene oxide spacer chain, with x=1,3,7 and 12, reported that the hydrophilic spacer with oxyethylene moieties was not fully extended at the air-water interface. With increasing the spacer group size x = 7-10, it became sufficiently flexible to adopt a particular conformation with a loop at the water side of the interface (Denjeriou-Sedjerari *et al.*, 2009). In addition, as observed for the 12-s-12 gemini surfactants, the micellization and binding process seems to be entropy driven for the 12-EOx-12 surfactants. In case of 12-8N-12 surfactant, the aza groups represent a bulkier, and perhaps more importantly, a more hydrophobic substituent as compared to oxygen for the case of the ethoxylated spacer groups. This has two direct effects on the ability of the spacer group to adopt conformations that would minimize the conformation energies:

i. steric hindrance between the aza methyl groups and the alkyl tails will not allow the spacer group to fold into the core of the micelle; and

ii. steric considerations also require that at least one of the aza methyl groups be oriented towards the aqueous phase, resulting in a net decrease in the energy associated with the transfer of the spacer from the aqueous to the micellar phase.

The increase in CMC for the N-methyl subtituted series compared to the epoxy series results from increased steric repulsion, particularly with respect to packing at the surface of the micelle, and the interference

Table 1. Thermodynamic properties for the titration of 0.5 mM Gemini surfactant with DNA.[a]Data from Wettig et al., 2010.The binding of cationic alkyammonium surfactants to DNA was studied by Matulis et al., who broke the interaction of the surfactants with DNA down into an electrostatic ($\Delta_{elec}H$) and a hydrophobic contribution ($\Delta_{h\Phi}H$; Equation 2) (Matulis *et al.*, 2002).

Surfactant	ΔH_{init} (kJ mol^{-1})	P^-/N^+	ΔH_1 (kJ mol^{-1})	P^-/N^+	ΔH_2 (kJ mol^{-1})
12-2-12	10.2	0.24	26.3	0.41	28.7
12-EO$_1$-12	10.5	0.53	23.8	0.80	22.0
12-EO$_2$-12	11.2	0.46	15.0	0.61	15.5
12-EO$_1$-12	11.5	0.56	15.8	0.83	17.0
12-5N-12	13.8	0.38	21.5	0.69	20.1
12-8N-12	17.7	0.31	25.0	0.60	26.1
12-4-12	6.8	0.27	20.5	0.50	24.7
12-4(OH)-12	5.8	0.24	17.5	0.37	21.4
12-4(OH)$_2$-12	4.0	0.30	18.7	0.48	19.0
12-3-12[a]	4.7	0.41	25.1	0.73	24.2
12-EO$_3$-12	12.1	0.42	18.9	0.68	17.6

between the solvation shells around nitrogen and hydrocarbon regions within the spacer (Wettig *et al.*, 2007). A hydrophilic, flexible spacer prompts micelle formation, which leads to a smaller CMC, smaller alpha, larger N, and more negative ΔG_{mic} (Wang *et al.*, 2004). One would expect that the sources of contributions to the enthalpy of binding remain the same within the series. Since the length and composition of the alkyl tails is kept constant, the contribution to ΔH_M arising from the transfer of the alkyl tails from the bulk to the micelle should also remain constant. Therefore, the observed differences between the gemini surfactants with and without EO/aza/OH groups in the spacer must be associated with differences in hydration of the spacers of these surfactant series, when they are monomers in the bulk phase or located in the micelle/water/solution interface and when they are interacting with DNA. The initial enthalpies (ΔH_{init}) and the enthalpies at the maximum of the main peaks observed in Figure 6 (ΔH_1 and ΔH_2), along with their positions (in terms of charge ratio (P-/N+) are reported in Table 1. It is interesting to note that the difference between the initial enthalpy (upon titration of the gemini surfactants with DNA) and the enthalpy at the first peak observed inFigure 6 (ΔH_{init} - ΔH_1) is comparable to the enthalpies of micellization for the pure

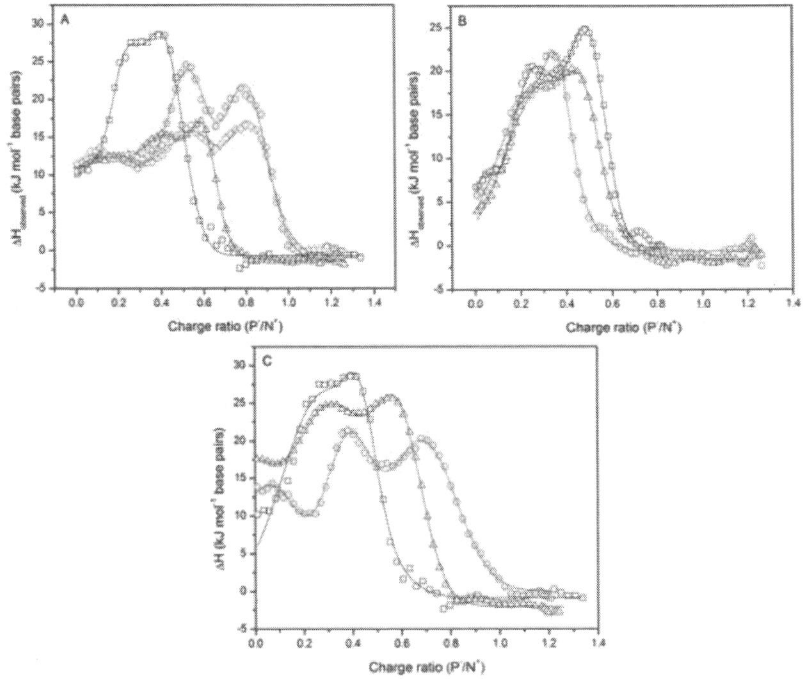

FIGURE 6. Observed enthalpies for the titration of gemini surfactant solutions with DNA: A) 12-2-12 (, black), 12-EO$_1$-12 (, red), 12-EO$_2$-12 (, blue) 12-EO$_3$-12 (, green); B) 12-4-12 (, black), 12-4(OH)$_1$-12 (, red), 12-4(OH)$_2$-12 (, blue); C) 12-2-12 (, black), 12-5N-12 (, red), 12-8N-12 (, blue). Lines represent a fit of the experimental data to a multiple Gaussian peak model.

surfactant compounds (Figure 7, and also ΔH_{mic} in Table 2) with a correlation of 0.72. As discussed above, one of the main differences in the mechanism proposed in our 2010 article (Wettig et al., 2010) and that originally proposed in our 2007 article (Wang et al, 2007) is the inclusion of the monomer-micelle equilibrium (Box C in Scheme 3). The monomer-micelle equilibrium is of course governed by the molecular structure of the surfactants, and therefore it is perhaps not surprising that we observe a correlation between the enthalpies of micellization and those observed upon interaction with DNA.

$$\Delta H = \Delta_{h\Phi}H + \Delta_{elec}H \qquad (2)$$

The electrostatic component was estimated to be ~1.0 kJ/mol (obtained from the enthalpy of dissolution for ammonium phosphate which allowed for the determination of the hydrophobic contribution from experimental enthalpy measurements. Enthalpy titrations for the alkylammonium surfactant – DNA systems were carried out in the usual (forward) manner of adding concentrated surfactant solution to a solution of DNA, rather than the reverse manner of adding DNA to a concentrated surfactant solution used in our work above. This unfortunately precludes a more detailed analysis and comparison of our results for the substituted gemini surfactants to those for the alkylammonium surfactants (Matulis *et al.*, 2002) as well as the gemini surfactants (Bai *et al.*, 2001;Jiang *et al.*, 2005;Wang *et al.*, 2007; Pozharski and MacDonald, 2002). Forward titrations with the substituted compounds are currently underway to allow for such a comparison.

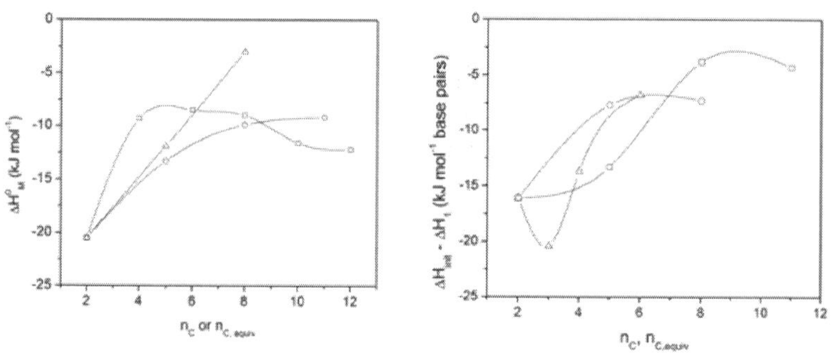

FIGURE 7. Enthalpies of micellization (ΔH_{mic}) for the 12-s-12 the 12-EO$_x$-12 and 12-Az$_X$-12 gemini surfactants; and the difference in enthaply at peak 1 and the initial enthalpy (ΔH_{init} - ΔH_1) as a function of spacer length for the titration of DNA with the 12-s-12, the 12-EO$_x$-12 and 12-Az$_X$-12 gemini surfactants.

The advantage of the reverse titration method is that it allows for an examination of complex formation under conditions of excess lipid (Kennedy *et al.*, 2000), potentially revealing differences in binding of DNA by lipid or surfactant aggregates (vesicles, micelles, etc.) rather than with monomer lipid or surfactant molecules that would be studied in the forward titration manner. Significant differences in how the resulting complexes assemble have been noted in the comparison of forward and reverse titration methods for the EDOPC – DNA systems introduced in section 2.3 (Kennedy *et al.*, 2000), indicating the potential to control the types, or more specifically structure, of lipid-DNA complexes simply by controlling the order in which the components are added to each other. Such an explanation likely explains the very common observation that transfection efficiencies are highly dependent upon the mixing order of the various components.

Table 2. Enthalpies of micellization for aqueous gemini surfactants and the difference in the observed peak and initial enthalpies (ΔH_{init} - ΔH_1) for the titration of gemini surfactants with DNA. [a] Data from (Wettig *et al.*, 2002); [b]Data from Wettig et al., 2010.

Surfactant	ΔH_{mic} (kJ mol^{-1})	ΔH_{init} - ΔH_1 (kJ mol^{-1})
12-2-12	-20.5	-16.1
12-EO_1-12	-13.3	-13.3
12-EO_2-12	-9.9	-3.8
12-EO_3-12	-9.2	-4.3
12-5N-12	-11.9	-7.7
12-8N-12	-3	-7.3
12-4-12	-9.2[a]	-13.7
12-4(OH)-12	-12.1[a]	-11.7
12-4$(OH)_2$-12	-11.5[a]	-14.3
12-3-12	-19.3[b]	-16.1
12-6-12	-8.5[b]	-6.8

4. CONCLUSION

Isothermal microcalorimetry plays an important role in elucidating the binding mechanism of DNA with surfactants/ lipids used in non-viral gene therapy. Measurement of heats of interaction allows accurate determination of binding constants, reaction stoichiometry (n), enthalpy (ΔH) and entropy (ΔS), thereby providing a complete thermodynamic profile of the molecular interaction in a single experiment by employing an appropriate model. Structure-activity studies suggest that biophysical properties, such as size, charge, and morphology of the resulting DNA/lipid complexes determine transfection efficiency within one class of vector.

Several parameters effect transfection activity, such as structural variations of the cationic lipids (length and degree of unsaturation of the alkyl chains), the polarity of head groups, the structure of spacer, and the counterions. It is generally observed that the chemical arrangement of gemini surfactants provide a rather rich array of aggregate morphologies and solution properties that are dependent upon the nature and size of the linking group. More recently, ITC has been used to study the interaction of DNA with transfection systems based upon a family of dimeric surfactants or gemini surfactants, which consist of two traditional surface active groups linked chemically by a spacer group. ITC data shows whether the interaction between DNA and gemini surfactants is endothermic or exothermic and the observed enthalpy *vs.* charge ratio profile depends upon the titration sequence. The magnitude and sign (i.e., endo- or exo-thermic) of ΔH_{mic} is governed by the types and magnitude of various contributions (hydrophobic and electrostatic) arising from surfactant-surfactant interactions. Since, for any process to be thermodynamically favoured the free energy change must be negative, an endothermic lipoplex formation must be driven by a corresponding increase in entropy. It is widely accepted that the major contribution to this increase in entropy arises from counterion release. The double-peak feature observed in the enthalpogram for the titration of the 12-3-12 surfactant with DNA is common to all of the symmetric gemini surfactants investigated (so far) by our group. Our group is using ITC to better understand the thermodynamic properties of the complexation of DNA by the aza-, hydroxyl- and ethoxyl-gemini substituted and unsubstituted surfactants, specifically to examine how the binding interactions change as a result of substitution within the spacer group. In addition, our group is interested in the study of a non-symmetric gemini surfactant in which one of the alkyl tails is a dodecyl group (i.e. m = 12) and the other has been replaced with a (pyren-6-yl)-hexyl group which has the same approximate length as a dodecyl group, but is significantly more hydrophobic in character. The major difference observed in the binding interaction(s) with DNA for the pyrenyl gemini surfactants lies in the ability of the pyrenyl group to intercalated between DNA base pairs. Interestingly a simplification of the system is observed for the pyrenyl surfactants (py-3-12; py-6-12), such that a number of intermediate equilibria do not seem to occur. One of the main differences in the mechanism proposed in our 2010 article (Wettig et al., 2010) and that originally proposed in our 2007 article (Wang et al, 2007) is the inclusion of the monomer-micelle equilibrium which in turn is governed by the molecular structure of the surfactants. The significance of these results remains unclear; however, forward titrations of the substituted gemini surfactants with are underway. Ultimately these results will allow for a more complete picture of the importance of head group structure of the gemini surfactants for improved DNA transfection efficiency, and the rational design of improved non-viral transfection vectors.

REFERENCES

1. G. Bai, Y. Wang, H. Yan, R. K. Thomas, 2001 Enthalpies of Micellization of Double Chain and Gemini Cationic Surfactants. J Colloid Interface Sci 240 375 377 .
2. A. Bajaj, B. Paul, P. Kondaiah, S. Bhattacharya, 2008 Structure-activity investigation on the gene transfection properties of cardiolipin mimicking gemini lipid analogues. Bioconjug Chem 19 1283 1300 .
3. P. C. A. Barreleiro, G. Olofsson, P. Alexandridis, 2000 Interaction of DNA with cationic vesicles: A calorimetric study. J Phys Chem B 104 7795 7802
4. B. Barteau, R. Chevre, E. Letrou-Bonneval, R. Labas, O. Lambert, B. Pitard, 2008 Physicochemical parameters of non-viral vectors that govern transfection efficiency. Curr Gene Ther 8 313 323 .
5. C. Bombelli, F. Faggioli, P. Luciani, G. Mancini, M. G. Sacco, 2005 Efficient transfection of DNA by liposomes formulated with cationic gemini amphiphiles. J Med Chem 48 5378 5382 .
6. R. Bruinsma, 1998 Electrostatics of DNA-cationic lipid complexes: isoelectric instability. Eur Phys J B 4 75 88 .
7. E. Calvet, H. Prat, 1963 Recent progress in microcalorimetry. The MacMillan Company, New York.
8. M. Castro, D. Griffiths, A. Patel, N. Pattrick, C. Kitson, M. Ladlow, 2004 Effect of chain length on transfection properties of spermine-based gemini surfactants. Org Biomol Chem 2 2814 2820 .
9. V. Cherezov, H. Qiu, V. Pector, M. Vandenbranden, J. M. Ruysschaert, M. Caffrey, 2002 Biophysical and transfection studies of the diC(14)-amidine/DNA complex. Biophys J 82 3105 3117 .
10. A. Djeriou-Sedjerari, G. Derrien, C. Charnay, J. Zajac, L. C. De Menorval, M. Lindheimer, 2009 Contribution of 1H NMR to the investigation of the adsorption of cationic Gemini surfactants with

oligooxyethylene spacer group onto silica. J Colloid Interface Sci 331 281 287 .

11. M. Donkuru, I. Badea, S. Wettig, R. Verrall, M. Elsabahy, M. Foldvari, 2010 Advancing nonviral gene delivery: lipid- and surfactant-based nanoparticle design strategies. Nanomedicine (Lond) 5 1103 1127 .

12. A. Elouahabi, J. M. Ruysschaert, 2005 Formation and intracellular trafficking of lipoplexes and polyplexes. Mol Ther 11 336 347 .

13. K. K. Ewert, C. E. Samuel, C. R. Safinya, 2008 Lipid- DNA Interactions: Structure- Function Studies of Nanomaterials for Gene Delivery. In: DNA Interactions with Polymers and Surfactants, R Dias & B Lindman, (Eds), John Wiley & Sons, 978-0-470-25818-7, Nj, USA: 377 404 .

14. P. L. Felgner, T. R. Gadek, M. Holm, R. Roman, H. W. Chan, M. Wenz, et al. 1987 Lipofection: a highly efficient, lipid-mediated DNA-transfection procedure. Proc Natl Acad Sci U S A 84 7413 7417 .

15. P. L. Felgner, G. M. Ringold, 1989 Cationic liposome-mediated transfection. Nature 337 387 388 .

16. M. Foldvari, I. Badea, S. Wettig, R. Verrall, M. Bagonluri, 2006 Structural characterization of novel gemini non-viral DNA delivery systems for cutaneous gene delivery. Journal of Exp. Nanosci. 1 165 176

17. S. Giatrellis, G. Nikolopoulos, Z. Sideratou, G. Nounesis, 2009 Calorimetric study of the interaction of binary DMTAP/DOTAP cationic liposomes with plasmid DNA. J Liposome Res 19 220 230 .

18. E. D. Goddard, 2002 Polymer/Surfactant Interaction: Interfacial Aspects. J Colloid Interface Sci 256 228 235 .

19. L. Grosmaire, M. Chorro, C. Chorro, S. Partyka, R. Zana, 2002 Alkanediyl-alpha, omega-bis(dimethylalkylammonium bromide) surfactants 9. Effect of the spacer carbon number and temperature on the enthalpy of micellization. J Colloid Interface Sci 246 175 181 .

20. D. Hirsch-Lerner, Y. Barenholz, 1999 Hydration of lipoplexes commonly used in gene delivery: follow-up by laurdan fluorescence changes and quantification by differential scanning calorimetry. Biochim Biophys Acta 1461 47 57 .

21. S. W. Hui, M. Langner, Y. L. Zhao, P. Ross, E. Hurley, K. Chan, 1996 The role of helper lipids in cationic liposome-mediated gene transfer. Biophys J 71 590 599 .

22. N. Jiang, P. Li, Y. Wang, J. Wang, H. Yan, R. Thomas, 2004 Microcalorimetric Micellization of cationic Gemini Surfactants with various counterions ans their interaction with DNA in aqueous solution. Journal of Physical Chemistry 108 15385 15391 .

23. N. Jiang, J. Wang, Y. Wang, H. Yan, R. K. Thomas, 2005 Microcalorimetric study on the interaction of dissymmetric gemini surfactants with DNA. J Colloid Interface Sci 284 759 764 .

24. C. Kawaura, A. Noguchi, T. Furuno, M. Nakanishi, 1998 Atomic force microscopy for studying gene transfection mediated by cationic liposomes with a cationic cholesterol derivative. FEBS Lett 421 69 72 .

25. M. T. Kennedy, E. V. Pozharski, V. A. Rakhmanova, Donald. R. C. Mac, 2000 Factors governing the assembly of cationic phospholipid-DNA complexes. Biophys J 78 1620 1633 .

26. A. J. Kirby, P. Camilleri, J. B. Engberts, M. C. Feiters, R. J. Nolte, O. Soderman, et al. 2003 Gemini surfactants: New synthetic vectors for gene transfection. Angewandte Chemie-International Edition 42 11448 11457 .

27. I. Koltover, T. Salditt, J. O. Radler, C. R. Safinya, 1998 An inverted hexagonal phase of cationic liposome-DNA complexes related to DNA release and delivery. Science 281 78 81 .

28. R. Koynova, Y. S. Tarahovsky, L. Wang, Donald. R. C. Mac, 2007 Lipoplex formulation of superior efficacy exhibits high surface activity and fusogenicity, and readily releases DNA. Biochim Biophys Acta 1768 375 386 .

29. R. Koynova, L. Wang, Donald. R. C. Mac, 2008 Cationic phospholipids forming cubic phases: lipoplex structure and transfection efficiency. Mol Pharm 5 739 744 .

30. R. Koynova, B. Tenchov, 2009 Cationic phospholipids: structure-transfection activity relationships. Soft Matter 5 3187 3200 .

31. P. Kreiss, B. Cameron, R. Rangara, P. Mailhe, O. guerre-Charriol, M. Airiau, et al. 1999 Plasmid DNA size does not affect the physicochemical properties of lipoplexes but modulates gene transfer efficiency. Nucleic Acids Res 27 3792 3798 .

32. B. A. Lobo, S. A. Rogers, S. Choosakoonkriang, J. G. Smith, G. Koe, C. R. Middaugh, 2002 Differential scanning calorimetric studies of the thermal stability of plasmid DNA complexed with cationic lipids and polymers. J Pharm Sci 91 454 466 .

33. B. , S. Zhang, H. Jiang, B. Zhao, H. Lv, 2007 Lipoplex morphologies and their influences on transfection efficiency in gene delivery. J Control Release ;123 184 194 .

34. R. W. Malone, P. L. Felgner, I. M. Verma, 1989 Cationic liposome-mediated RNA transfection. Proc Natl Acad Sci U S A 86 6077 6081 .

35. D. Matulis, I. Rouzina, V. A. Bloomfield, 2002 Thermodynamics of cationic lipid binding to DNA and DNA condensation: roles of electrostatics and hydrophobicity. J Am Chem Soc 124 7331 7342 .

36. O'Hern, C.S. and Lubensky, T.C 1998 Sliding Columnar Phase of DNA-Lipid Complexes. Physical Review Letters 80 4345 4348 .

37. A. Pathak, S. Patnaik, K. C. Gupta, 2009 Recent trends in non-viral vector-mediated gene delivery. Biotechnol J 4 1559 1572 .

38. V. Pector, J. Backmann, D. Maes, M. Vandenbranden, J. M. Ruysschaert, 2000 Biophysical and structural properties of DNA.diC(14)-amidine complexes. Influence of the DNA/lipid ratio. J Biol Chem 275 29533 29538 .

39. M. M. Pierce, C. S. Raman, B. T. Nall, 1999 Isothermal titration calorimetry of protein-protein interactions. Methods 19 213 221 .

40. B. Pitard, N. Oudrhiri, J. P. Vigneron, M. Hauchecorne, O. Aguerre, R. Toury, et al. 1999 Structural characteristics of supramolecular assemblies formed by guanidinium-cholesterol reagents for gene transfection. Proc Natl Acad Sci U S A 96 2621 2626 .

41. E. Pozharski, Donald. R. C. Mac, 2002 Thermodynamics of cationic lipid-DNA complex formation as studied by isothermal titration calorimetry. Biophys J 83 556 565 .

42. E. Pozharski, Donald. R. C. Mac, 2003 Lipoplex thermodynamics: determination of DNA-cationic lipoid interaction energies. Biophys J 85 3969 3978 .

43. J. O. Radler, I. Koltover, T. Salditt, C. R. Safinya, 1997 Structure of DNA-cationic liposome complexes: DNA intercalation in multilamellar membranes in distinct interhelical packing regimes. Science 275 810 814 .

44. C. R. Safinya, 2001 Structures of lipid-DNA complexes: supramolecular assembly and gene delivery. Curr Opin Struct Biol 11 440 448 .

45. J. Smisterova, A. Wagenaar, M. C. Stuart, E. Polushkin, B. G. ten, R. Hulst, et al. 2001 Molecular shape of the cationic lipid controls the

structure of cationic lipid/dioleylphosphatidylethanolamine-DNA complexes and the efficiency of gene delivery. J Biol Chem 276 47615 47622 .

46. B. Sternberg, K. Hong, W. Zheng, D. Papahadjopoulos, 1998 Ultrastructural characterization of cationic liposome-DNA complexes showing enhanced stability in serum and high transfection activity in vivo. Biochim Biophys Acta 1375 23 35 .

47. B. Sternberg, F. L. Sorgi, L. Huang, 1994 New structures in complex formation between DNA and cationic liposomes visualized by freeze-fracture electron microscopy. FEBS Lett 19;356 361 366 .

48. K. Takeuchi, M. Ishihara, C. Kawaura, M. Noji, T. Furuno, M. Nakanishi, 1996 Effect of zeta potential of cationic liposomes containing cationic cholesterol derivatives on gene transfection. FEBS Lett 397 207 209 .

49. I. Tranchant, B. Thompson, C. Nicolazzi, N. Mignet, D. Scherman, 2004 Physicochemical optimisation of plasmid delivery by cationic lipids. J Gene Med 6 Suppl 1: S24 -S35.

50. C. Wang, X. Li, S. D. Wettig, I. Badea, M. Foldvari, R. E. Verrall, 2007a Investigation of complexes formed by interaction of cationic gemini surfactants with deoxyribonucleic acid. Phys Chem Chem Phys 9 1616 1628 .

51. C. Wang, S. D. Wettig, M. Foldvari, R. E. Verrall, 2007b Synthesis, characterization, and use of asymmetric pyrenyl-gemini surfactants as emissive components in DNA-lipoplex systems. Langmuir 23 8995 9001 .

52. X. Wang, J. Wang, Y. Wang, H. Yan, P. Li, R. K. Thomas, 2004 Effect of the nature of the spacer on the aggregation properties of gemini surfactants in an aqueous solution. Langmuir 20 53 56 .

53. S. D. Wettig, R. Deubry, J. Akbar, T. Kaur, H. Wang, T. Sheinin, et al. 2010 Thermodynamic investigation of the binding of dissymmetric pyrenyl-gemini surfactants to DNA. Phys Chem Chem Phys 12 4821 4826 .

54. S. D. Wettig, X. F. Li, R. E. Verrall, 2003 Thermodynamic aggregation properties of gemini surfactants with ethoxylated spacers in aqueous solution. Langmuir 19 3666 3670 .

55. S. D. Wettig, P. Nowak, R. E. Verrall, 2002 Aggregation Behavior of Hydroxyl Substituted Gemini Surfactants in Aqueous Solution. Langmuir 18 5354 5359 .

56. S. D. Wettig, C. Wang, R. E. Verrall, M. Foldvari, 2007 Thermodynamic and aggregation properties of aza- and imino-substituted gemini surfactants designed for gene delivery. Phys Chem Chem Phys 9 871 877 .

57. P. Yang, J. Singh, S. Wettig, M. Foldvari, R. E. Verrall, I. Badea, 2010 Enhanced gene expression in epithelial cells transfected with amino acid-substituted gemini nanoparticles. Eur J Pharm Biopharm 75 311 320 .

58. J. Zabner, A. J. Fasbender, T. Moninger, K. A. Poellinger, M. J. Welsh, 1995 Cellular and molecular barriers to gene transfer by a cationic lipid. J Biol Chem 270 18997 19007 .

59. J. S. Zhang, S. Li, L. Huang, 2003 Cationic liposome-protamine-DNA complexes for gene delivery. Methods Enzymol 373 332 342 .

60. S. Zuzzi, C. Cametti, G. Onori, S. Sennato, 2007 Liposome-induced DNA compaction and reentrant condensation investigated by dielectric relaxation spectroscopy and dynamic light scattering techniques. Phys Rev E Stat Nonlin Soft Matter Phys 76: 011925.

6

CHAPTER

Thermodynamics for trajectories of a mass point

Yoshimasa Kurihara, Khiem Hong Phan, Nhi My Uyen Quach

The High Energy Accelerator Organization (KEK)

The Graduate University for Advanced Studies

ABSTRACT

On the basis of the theory of thermodynamics, a new formalism of classical nonrelativistic mechanics of a mass point is proposed. The particle trajectories of a general dynamical system defined on a (1+n)(1+n)-dimensional smooth manifold are geometrically treated as dynamical variables. The statistical mechanics of particle trajectories are constructed in a classical manner. Thermodynamic variables are introduced through a partition function based on a canonical ensemble of trajectories. Within this theoretical framework, classical mechanics can be interpreted as an equilibrium state of statistical mechanics. The relationship between classical and quantum mechanics is discussed from the viewpoint of statistical mechanics. The maximum-entropy principle is shown to provide a unified view of both classical and quantum mechanics.

INTRODUCTION

Quantum mechanics is considered to be the most basic theory of nature. All phenomena, including gravitational interactions, have an underlying quantum-mechanical interpretation. Quantum mechanics describes the microscopic behavior of particles under fundamental forces and has been adopted in numerous applications. However, our understanding of quantum mechanics remains incomplete. One of the most characteristic and mysterious aspects of quantum mechanics is that particle properties are described by probability amplitudes. The probabilistic aspects of quantum mechanics are inherent characteristics and are not due to lack of detailed information such as partition functions in statistical mechanics. Therefore, understanding why and how the probabilistic nature of quantum mechanics emerges from a primary principle is of critical importance. To pursue this purpose, we propose herein to use a thermodynamic theory.

Let us recall the relationship between thermodynamics and statistical mechanics. Thermodynamics is a field of physics that discusses the relationship between the macroscopic physical quantities such as temperature, pressure, volume, energy, entropy, and heat and/or work from outside of the system. Thermodynamics was established before the microscopic details were clarified. Later, statistical mechanics was constructed on the basis of the microscopic details of classical mechanics using thermodynamics as a guiding principle. However, statistical mechanics based on classical mechanics failed to explain, for instance, the entire nature of electromagnetic waves radiated from gases and metals, which provided a hint about quantum mechanics. Although the microscopic details of statistical mechanics were replaced by quantum mechanics instead of by classical mechanics, the consequences of thermal dynamics remain true, and again thermal dynamics plays the role of a guiding principle in constructing a theory. Thermodynamics and its second law are still active in the field of physics, and are discussed, for instance, by Lieb and Yngvason [12] in their epoch-making paper. Relationships between thermodynamics and quantum mechanics are also intensively discussed by Gemmer et al. [6]. They gave new explanations for the emergence of thermodynamics behavior from a quantum mechanical system. Our way is something opposite to their intention, "the emergence of quantum mechanics from the thermodynamics".

In our pursuit of the principle that underlies the probabilistic characteristics of the basic equations of motion, we suspend the notion that quantum mechanics is the most fundamental theory of nature and regard it as a phenomenological theory. In other words, we propose to construct the thermodynamics of quantum mechanics and then pursue the underlying mechanics, which must be a more fundamental theory of nature. In this study, we attempt to construct the thermodynamics of classical and quantum mechanics of a mass point. To construct the thermodynamics of classical mechanics, an appropriate definition of entropy must be introduced. To determine the entropy of the classical dynamical system of a mass point, we consider the system in geometrical terms and introduce a general dynamical system a′laa′laArnol'd et al. [1]. Then, the analogy between the thermodynamics of gases and the Hamiltonian formalism of classical dynamics guides us to identify an appropriate definition of entropy. An equation of motion of the system can be extracted by requiring the maximum-entropy principle (instead of the principle of least action). Finally, quantum fluctuations can be understood in analogy with thermal fluctuations. This view may bring new insights of quantum mechanics.

This paper is organized as follows: After introducing a general framework for dynamic systems in Sect. 1, a generalized Hamiltonian formalism is developed in Sect. 2. The statistical mechanics of particle trajectory in the proposed framework is developed in Sect. 3. Here, we demonstrate that a thermal-equilibrium state of the trajectory corresponds to the classical mechanics of a mass point. Section 4 is devoted to relating classical and quantum mechanics using the statistical mechanics analogy. Conclusions and further discussion are presented in Sect. 5.

GENERAL MECHANICS

Dynamics on a symplectic manifold

First, we introduce a general dynamic space on which various dynamical systems are developed. A structure given here is a common for classical dynamic systems treated in this report. It is considered to give the minimum mathematical system to discuss classical mechanics.

Definition 2.1

General Dynamic Space A general dynamical space is a collection of sets, { $\{\mathcal{M} = \mathscr{M} \otimes \mathscr{T}, f, \mathcal{P} = \boldsymbol{\xi} \otimes \eta, (\omega, \Omega)\}$ whose elements are defined as follows:

1. A manifold $\mathscr{M} \in \mathbb{R}^n$ is an nn-dimensional Euclidean space called a space manifold.

2. A manifold $\mathscr{T} \in \mathbb{R}$ is a one-dimensional smooth manifold called a time manifold. A point on TT is called an order parameter or time.

3. The direct product $\mathcal{M} = \mathscr{T} \otimes \mathscr{M}$ is a space-time manifold. A position vector ξξξξ on an open neighborhood UpUp of a point pp can be expressed in terms of local orthonormal bases as

$$\xi = (\xi^0 = \tau, \xi^1, \cdots, \xi^n).$$

A flat metric whose metric tensor gμμ=(−,+,...,+)gμμ=(−,+,...,+) is imposed on the space-time manifold. Then, the MM becomes a Riemannian manifold with an indefinite metric.

4. A tangent bundle of MM is written as $T\mathcal{M} = \bigcup_{p\in\mathcal{M}} T_p\mathcal{M}.$ tangent vector to ξξξξ is expressed as.

$$\partial_\xi = \left(\frac{\partial}{\partial\xi^0} = \frac{\partial}{\partial\tau}, \frac{\partial}{\partial\xi^1}, \cdots, \frac{\partial}{\partial\xi^n}\right) \in T\mathcal{M}.$$

5. In the same manner, we introduce a cotangent bundle, $T^*\mathcal{M},$, with cotangent vector to ξ expressed as

$$d\xi = \left(d\xi^0 = d\tau, d\xi^1, \cdots, d\xi^n\right) \in T^*\mathcal{M}.$$

6. A characteristic function f is a C^∞- function that maps a point on M to a real number. The characteristic function is assumed holomorphic for a position vector ξ such that $\partial^2 f/\partial\xi^\mu\partial\xi^\nu = \partial^2 f/\partial\xi^\nu\partial\xi^\mu,\ \mu,\nu = 0,\cdots,n.$

7. A momentum vector is introduced in terms of the characteristic function as

$$\begin{aligned}\eta &= (\eta^0, \cdots, \eta^n), \\ &= \partial_{\xi^*} f = \left(\frac{\partial f}{\partial\xi_0}, \cdots, \frac{\partial f}{\partial\xi_n}\right),\end{aligned}$$

where $\xi^* = \{g_{\mu\nu}\xi^\mu\}.$ Here we follow the Einstein convention in summing the repeated indices, summing the Greek indices from 00 to nn, and summing the Roman indices from 11 to nn, unless otherwise stated. Not all η's and ξ's are independent because the characteristic function imposes a constraint. We assume that the zeroth component of the momentum vector is a function of other components, $\eta^0(\eta^1, \cdots, \eta^n, \xi^0, \cdots, \xi^n).$ A direct product of position and momentum vector space $\mathcal{P} = \xi \otimes \eta$ is called an extended phase space. This space is a (2n+2)-dimensional smooth manifold.

8. On the extended phase-space, 11- and 22-forms such as

$$\omega = \eta \cdot d\xi$$
$$= \eta_\mu d\xi^\mu \quad (1)$$
$$= \eta_i d\xi^i - \eta_0 d\tau, \quad (2)$$

$$\Omega = d\omega$$
$$= d\eta_\mu \wedge d\xi^\mu \quad (3)$$
$$= d\eta_i \wedge d\xi^i - d\eta_0 \wedge d\tau,$$

are defined [1], which are called characteristic 11-form and 22-form, respectively. The characteristic 22-form is assumed to be a closed form satisfying d Ω=0. The even dimensional manifold that has a non-degenerate and closed 22-form is called a symplectic manifold.

Definition 2.2

Time evolution operator and Trajectory The smooth map

$$\phi_s : \mathcal{M} \to \mathcal{M} : \xi = (\tau, \xi^1, \cdots, \xi^n) \mapsto \phi_s \xi = \xi' = (\tau + s, \xi'^1(\tau + s), \cdots, \xi'^n(\tau + s)),$$

is called a time evolution operator. The map ϕsϕs generates a one-dimensional manifold γ⊂M, such that

$$\gamma = \{\xi(\tau) = (\tau, \xi^1, \cdots, \xi^n)(\tau) \mid \phi_s(\tau_1, \xi'^1(\tau_1 + s), \cdots, \xi'^n(\tau_1 + s), s \in [0, \tau_2 - \tau_1]\}.$$

This manifold is called a trajectory.

We assume that a tangent vector exists for a given trajectory, i.e.,

$$\left.\frac{d\xi(\tau)}{d\tau}\right|_{\tau=t} = \lim_{\delta t \to \pm 0} \frac{\xi(t + \delta t) - \xi(t)}{\delta t},$$

at any $t \in [\tau_1, \tau_2]$. The time evolution operator maps a momentum vector as

$$\eta' = \phi_s \eta = \partial_{\xi^*} f \Big|_{\xi = \phi_s \xi}.$$

The time evolution operator, which introduces dynamics to the general dynamic space, must describe some physical principle. More specifically, the following is true.

Principle 1

(Cartan) [2, 15] When the integration of the characteristic 11-form is invariant under map ϕsϕs, i.e.,

$$\int_l \omega(\xi,\eta) = \int_{l(\phi_s\xi,\phi_s\eta)} \omega(\phi_s\xi,\phi_s\eta),$$

the trajectory induced by ϕsϕs is physically realized. Here $l = l(\xi,\eta)$ is any closed circle in an extended phase space (P)(P) at fixed ττ.

Theorem 2.1

(Characteristic Equations) Trajectories satisfying Principle 1 determine an equation of motion such that

$$\frac{d\tilde{\xi}^i}{ds} = \frac{\partial\tilde{\eta}_0}{\partial\eta^i}, \quad \frac{d\tilde{\eta}_i}{ds} = -\frac{\partial\tilde{\eta}_0}{\partial\xi^i}, \quad \frac{d\tilde{\eta}_0}{ds} = \frac{\partial\tilde{\eta}_0}{\partial s},$$

which are known as characteristic equations.

Here $\tilde{\eta} = (\tilde{\eta}_0, \tilde{\eta}_1(s), \cdots, \tilde{\eta}_n(s))$ is the tangent vector along the trajectory $\tilde{\xi}(s)$, defined as

$$\tilde{\eta}(s) = \partial_{\xi^*} f\Big|_{\xi=\tilde{\xi}(s)},$$

where $\tilde{\eta}^0 = \partial f/\partial\tau|_{\tau=s}$ is assumed to be a function of ξ and η_i, $(i = 1, \cdots, n)$.

Proof

The characteristic forms along the trajectory ξ~(s)ξ~(s) can be written as

$$\begin{aligned}
\tilde{\omega} &= \tilde{\eta}_\mu d\tilde{\xi}^\mu \\
&= -\tilde{\eta}_0 d\tau + \tilde{\eta}_i d\tilde{\xi}^i, \\
\tilde{\Omega} &= d\tilde{\eta}_\mu \wedge d\tilde{\xi}^\mu \\
&= -d\tilde{\eta}_0 \wedge d\tilde{\xi}^0 + d\tilde{\eta}_i \wedge d\tilde{\xi}^i,
\end{aligned}$$

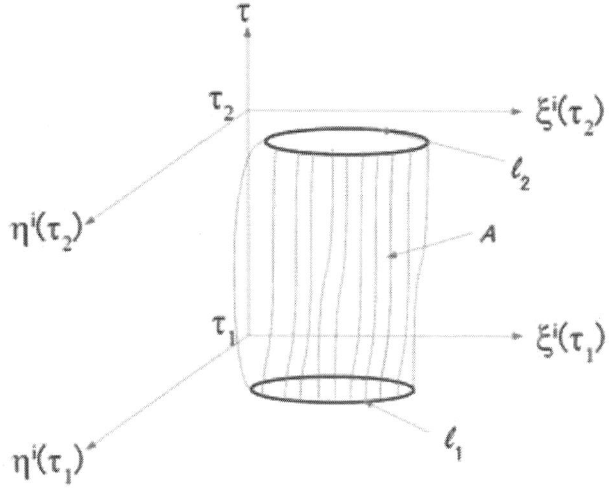

Figure. 1. Cartan tube

and by definition

$$\begin{aligned} d\tilde{\xi}^0 &= d\tau, \\ d\tilde{\eta}_0 &= \left[\frac{\partial\tilde{\eta}_0}{\partial\xi^i}d\xi^i + \frac{\partial\tilde{\eta}_0}{\partial\eta_i}d\eta_i + \frac{\partial\tilde{\eta}_0}{\partial\tau}d\tau\right]_{\tau=s}, \end{aligned} \quad (4)$$

$$= \frac{\partial\tilde{\eta}_0}{\partial\tilde{\xi}^i}d\tilde{\xi}^i + \frac{\partial\tilde{\eta}_0}{\partial\tilde{\eta}_i}d\tilde{\eta}_i + \frac{\partial\tilde{\eta}_0}{\partial s}ds. \quad (5)$$

Here we used a property of an exterior derivative, $d(\omega\wedge\eta) = d\omega + (-1)^{deg\ \omega}(\omega\wedge\ d\eta)$ for any ω and η, and $\partial\tilde{\xi}_0/\partial s = 1$. Note that Eq. (5) is a simplified expression of Eq. (4). Consider the *Cartan tube* shown in Fig. 1. Let l1 be a closed circle at $\tau = \tau_1$, and $l_2 = \phi_{\tau_2-\tau_1}(\xi \in l_1)$. The cylindrical surface of the Cartan tube AA consists of trajectories connected from l1 to l2. By Stokes' theorem, an integration of the characteristic 22-form on AA vanishes,1 i.e.,

$$
\begin{aligned}
\int_A \tilde{\Omega} &= \int_A d\tilde{\omega}, \\
&= \int_{\partial A} \tilde{\omega}, \\
&= \int_{l_2} \tilde{\omega} - \int_{l_1} \tilde{\omega}, \\
&= 0,
\end{aligned} \tag{6}
$$

where the last equivalence follows from the definition of the trajectory, which maintains constant ω~ω~. On the other hand, by contraction of the left-hand side of Eq. (6) with $\phi_s = ds\partial/\partial s$, we obtain

$$
\begin{aligned}
0 &= \int_A \langle \tilde{\Omega} | \phi_s \rangle, \\
&= \int_A \left(-\frac{d\tilde{\eta}_0}{ds} d\tilde{\xi}^0 + \frac{d\tilde{\eta}_i}{ds} d\tilde{\xi}^i - d\tilde{\eta}_0 \frac{d\tilde{\xi}^0}{ds} + d\tilde{\eta}_i \frac{d\tilde{\xi}_i}{ds} \right), \\
&= \int_A \left(\frac{\partial \tilde{\eta}_0}{\partial s} - \frac{d\tilde{\eta}_0}{ds} \right) ds + \int_A \left(\frac{d\tilde{\eta}_i}{ds} + \frac{\partial \tilde{\eta}_0}{\partial \xi^i} \right) d\tilde{\xi}^i + \int_A \left(\frac{\partial \tilde{\eta}_0}{\partial \eta^i} - \frac{d\tilde{\xi}_i}{ds} \right) d\tilde{\eta}_i,
\end{aligned} \tag{7}
$$

where ⟨·|·⟩ is a contraction of two forms. The third step of this contraction uses Eq. (5). For the integral to identically vanish on any AA, each term in the parentheses must be zero. □

This proof immediately leads to the following remark.

Remark 1

(Coordinate independent representation) The characteristic equations can be expressed independently of the coordinates as

$$\langle \tilde{\Omega} | \phi_s \rangle = 0.$$

Proof

The proof is evident from Eq. (7), which is true for all AA. □□

Hereafter, the ~~ of the trajectory is omitted for simplicity.

Definition 2.3

(Hamiltonian, Lagrangian and Action) A Hamiltonian, HH, is defined from the zeroth component of the momentum vector as

$$\mathcal{H} = \eta_0. \tag{8}$$

from which an integration of the characteristic 11-form along a trajectory can be expressed as

$$\begin{aligned}\mathscr{I}(\gamma) &= \int_\gamma \omega, \\ &\models \int_\gamma \left(\eta_i d\xi^i - \mathcal{H} d\tau\right), \\ &= \int_{\tau_0}^{\tau_1} d\tau \gamma^* \left(\eta_i \frac{d\xi^i}{d\tau} - \mathcal{H}\right).\end{aligned}$$

This integral is called an action, where γγ is the trajectory and $\gamma^*(\bullet)$ is a pull of (·) by γ. The characteristic11-form can be expressed as

$$\begin{aligned}\omega &= \left(\gamma^* \mathcal{L}\right)(\tau) d\tau, \\ \mathcal{L} &= \eta_i \frac{d\xi^i}{d\tau} - \mathcal{H}.\end{aligned} \quad (9)$$

Here L is the Lagrangian.

Since the action is independent of the parameterization of the trajectory, the pull of the trajectory γ∗ is hereafter omitted unless required for clarity. The action can be interpreted as the "distance" between two points on the space-time manifold measured by the characteristic 11-form ωω. A detailed treatment of the trajectory γγ is given in Sect. 3.

Examples of dynamics

Here we provide two concrete examples of the general dynamical space from thermodynamics and Hamiltonian mechanics. Similarity between these two examples guides us to new insights in classical dynamics.

Example 1

(Irreversible thermodynamics)2

First, let us consider an isolated system of gas. The adiabatic free expansion of an isolated system enlarges the entropy (S) (according to the second law of thermodynamics) and maintains constant temperature (T)because heat energy cannot be gained or lost. In this case, the order parameter is entropy (S∈T), while the one-dimensional space-manifold is volume (V∈M). The thermodynamics is described by the following characteristic function:

$$f_{TD} = pV - ST, \quad (10)$$

where pp is the pressure of gas in an insulating container and pp is the pressure of the system. From the thermodynamic characteristic function, η0η0 can be obtained as ∂fTD/∂S=T∂fTD/∂S=T. The general dynamical space then becomes

$$
\begin{aligned}
(\xi^0, \xi^1) &= (S, V), \\
(\eta^0, \eta^1) &= (T, p), \\
\omega_E &= pdV - TdS, \\
\Omega_E &= dp \wedge dV - dT \wedge dS.
\end{aligned}
$$

The characteristic function can be expressed as a scalar function on the space-time manifold as follows:

$$
\begin{aligned}
f_{TD} &= \xi^\mu \eta_\mu, \\
&= \xi^1 \eta_1 - \xi^0 \eta_0, \\
&= pV - ST.
\end{aligned}
$$

We assume that a Hamiltonian exists such that η0=T(p,V)η0=T(p,V). From the characteristic function and space-time manifold, we obtain the following characteristic equations:

$$
\frac{dV}{dS} = \frac{\partial T}{\partial p}, \quad \frac{dp}{dS} = -\frac{\partial T}{\partial V}, \quad \frac{dT}{dS} = \frac{\partial T}{\partial S} = 0.
$$

These are known as the Maxwell relations for an adiabatic transition. The characteristic 11-form corresponds to the energy of the system $(\omega_E = -dE)$ 3

Next let us consider a system contained in an insulating container with an expandable wall and its isothermal reversible transition. The initial temperature of the system is T1. The system is attached to a heat bath of temperature T2>T1 and pressure is maintained constant. Here temperature is adopted as the order parameter (T∈T) because temperature increases monotonically in this case. In this example, the one-dimensional space-manifold is pressure p∈M. From the characteristic function, Eq. (10), η0η0 can be obtained as ∂fTD/∂T=S(V,p)∂fTD/∂T=S(V,p), and the general dynamical space becomes

$$
\begin{aligned}
(\xi^0, \xi^1) &= (T, p), \\
(\eta^0, \eta^1) &= (S, V), \\
\omega_G &= V dp - S dT, \\
\Omega_G &= dV \wedge dp - dS \wedge dT,
\end{aligned}
$$

and the characteristic functions are derived as

$$
\frac{dp}{dT} = \frac{\partial S}{\partial V}, \quad \frac{dV}{dT} = -\frac{\partial S}{\partial p}, \quad \frac{dS}{dT} = \frac{\partial S}{\partial T} = 0.
$$

These functions are alternative expressions of the above-derived Maxwell relations. Here the characteristic 11-form corresponds to the Gibbs free energy of the system, given as

$$
\omega_G = V dp - T dS = dG. \tag{11}
$$

Example 2

(Hamiltonian formalism of mass points)

Next let us consider the well-known Hamiltonian formalism of mass points with nn degrees of freedom. The order parameter and space-time manifold are defined as $t \in \mathscr{T}$ and $(q^1, \cdots, q^n) \in \mathscr{M}$, respectively. The characteristic function is

$$
\begin{aligned}
f &= \xi^\mu \eta_\mu, \\
&= q^i p_i - Ht,
\end{aligned}
$$

where $\xi = (t, q^1, \ldots, q^n)$ and $\eta = (H, p_1 \cdots, p_n)$. Assuming the Hamiltonian as $H = H(t, q^i, p_i)$, the characteristic forms are obtained as

$$
\begin{aligned}
\omega &= p_i dq^i - H dt, \\
\Omega &= dp_i \wedge dq^i - dH \wedge dt.
\end{aligned}
$$

Then, the celebrated canonical equations of motion can be expressed as the following characteristic functions:

$$
\frac{dq^i}{dt} = \frac{\partial H}{\partial p_i}, \quad \frac{dp_i}{dt} = -\frac{\partial H}{\partial q_i}, \quad \frac{dH}{dt} = \frac{\partial H}{\partial t}.
$$

Underlying structure of Hamiltonian systems

The similarity between thermodynamics and the Hamiltonian formalism of mass points was highlighted in the previous section. Both systems show the same symplectic structure of base manifold and evolve along an order parameter under a set of "equations of motion". However, thermodynamic ideal gas systems are known to possess an underlying structure generated by the statistical mechanics of independent gas molecules. On the other hand, the Hamiltonian formalism of mass points requires no underlying structure. Here we treat the Hamiltonian formalism as a thermodynamic system and assume a virtual underlying structure for the motions of mass points. Among several candidates for a microscopic entity governing the Hamiltonian formalism, particle trajectories are adopted for the following reasons. The Hamiltonian formalism of mass points differs from ideal-gas thermodynamics primarily by importance of the trajectory. The main goal of the former is to determine the trajectory of a mass point under applied forces and initial conditions. In contrast, in the latter, the trajectory cannot be measured and has no essential meaning, similar to the quantum mechanics of mass points. By considering particle trajectories as the statistical entity, a relation between classical and quantum mechanics may be clarified. This section considers the statistical mechanics of particle trajectories in the general dynamic space of Definition 2.1.

Geometrical preparation

This subsection introduces the geometrical objects used in subsequent discussions.

Definition 3.1

(Curvilinear path) A set of maps γ such that

$$\begin{aligned}\gamma &: \mathscr{T} \to \Gamma \subset \mathcal{M} : t \in [t_1, t_2] \mapsto \gamma(t) = \left(\gamma^0(t) = t, \gamma^1(t), \gamma^2(t), \cdots, \gamma^n(t)\right), \\ \gamma^i &: \mathbb{R} \to \mathbb{R} : t \mapsto \gamma^i(t), \quad \gamma^i \in C^\infty,\end{aligned}$$

is called a curvilinear path (or simply "path"), and a set of paths, ΓΓ, is called a curvilinear-path space.

In this section, we consider only those paths whose end points are fixed at $\gamma(t_1) = \boldsymbol{\xi}_1 = (t_1, \xi_1^1, \cdots, \xi_1^n)$ and $\gamma(t_2) = \boldsymbol{\xi}_2 = (t_2, \xi_2^1, \cdots, \xi_2^n)$.

Definition 3.2

(Velocity vector) A velocity vector is a tangent vector at a point γ(t)γ(t) on the curvilinear path, defined as

$$\frac{d\gamma}{dt}(t) \quad = \quad \left(1, \frac{d\gamma^1}{dt}(t), \frac{d\gamma^2}{dt}(t), \cdots, \frac{d\gamma^n}{dt}(t)\right).$$

Hereafter, the velocity vector is written as γ˙(t)=(1,γ˙1(t),γ˙2(t),...,γ˙n(t))γ˙(t)=(1,γ˙1(t),γ˙2(t),...,γ˙n(t)). The velocity vector can be expressed in terms of natural bases on a tangent space $T_{t\gamma}$ at $\gamma_{(t)}$ as

$$\dot{\gamma}^i(t) = \frac{d\gamma^i(s)}{ds}\frac{\partial}{\partial\gamma^i}\bigg|_{s=t},$$

where i runs from 1 to nn and the components are not summed. A tangent vector bundle $\dot{\Gamma} = \bigcup_{\gamma(t)\subset\gamma\in\Gamma} T_t\gamma$ is called a velocity bundle.

Definition 3.3

(Variational vector) A variational vector is defined as a cotangent vector at $\gamma(t) \in \mathcal{M}$ on the curvilinear path $\delta\gamma(t)$. In terms of natural bases of the cotangent space $T_t^*\gamma$, variational vectors can be expressed as $\delta\gamma^i(t) = \delta\gamma^i d\gamma_i$, where ii runs from 11 to nn and the components are not summed. The zeroth component is $\delta\gamma^0(t) = 0$. A cotangent vector bundle $\delta\Gamma = \bigcup_{\gamma(t)\subset\gamma\in\Gamma} T_t^*\delta\gamma$ is called a variational bundle.

The bases of a velocity bundle and variational bundle are orthogonal, i.e., $d\gamma^\mu\partial/\partial\gamma^\nu = \delta^\mu_\nu$.

Definition 3.4

(Variational operator) A map induced by a variational vector such that

$$\delta : \Gamma \to \Gamma : \gamma \mapsto (\delta \circ \gamma)(t) = (\gamma + \delta\gamma)(t),$$ is called a variational operator.

Here $(\gamma + \delta\gamma)(t)$ denotes the sum of two vectors γγ and δγδγ, which are defined on MM. The curvilinear path $\delta_\gamma(t) = (\delta \circ \gamma)(t)$ is assumed to become an element of the curvilinear-path space, i.e., $\delta_\gamma(t) \in C^\infty$. The distance between two paths γγ and δγ is defined as

$$\begin{aligned} ||\delta_\gamma - \gamma|| &= ||\delta\gamma||, \\ &= \int_{t_1}^{t_2} dt|\delta\gamma(t)|, \\ &= \int_{t_1}^{t_2} dt \left\{ \sum_{k=1}^{n} \left(\delta\gamma^k(t)\right)^2 \right\}^{1/2}. \end{aligned}$$

Suppose that FF is a functional defined on ΓΓ such that $\mathscr{F} : \Gamma \to \mathbb{R}^m : \gamma \mapsto \mathscr{F}(\gamma),\ m \in \mathbb{N}.$ The variational operator maps a functional $\mathscr{F}$ to another functional as $\delta : \mathscr{F}(\gamma) \mapsto \delta\mathscr{F}(\gamma) = \mathscr{F}(\delta_\gamma).$ Since γγis a map defined in Definition 3.1, a functional $\mathscr{F}$ can be pulled back to a function defined on $\mathbb{R}$ using a pull of γ, which denoted as γ∗:

$$\gamma^* : \mathscr{F}(\gamma) \mapsto \gamma^*\left(\mathscr{F}(\gamma)\right)(\tau) \in \mathbb{R}.$$

For simplicity, we use a shorthand, $\gamma^*(\mathscr{F}(\gamma))(\tau) = \mathscr{F}(\gamma)(\tau).$ variational operator can be pulled as $\gamma^*(\delta_\gamma \mathscr{F}(\gamma))(\tau) = \mathscr{F}(\gamma + \delta\gamma)(\tau).$ Thus variation of a functional FF can be defined as

$$\frac{\delta \parallel \mathscr{F}(\gamma) \parallel}{\delta\gamma} = \lim_{\|\delta\gamma\| \to 0} \frac{\parallel \mathscr{F}(\gamma + \delta\gamma) \parallel - \parallel \mathscr{F}(\gamma) \parallel}{\parallel \delta\gamma \parallel}.$$

When a variation is zero with some γcγc, it is said that the functional FF has a extremal at γc.

Dynamics of paths

We now define the general dynamic space occupied by a mass point and impose a probability space on it. For simplicity, we treat a single mass point.

Definition 3.5

(*General dynamic path space*) The general dynamical space occupied by a point particle is described as follows:

- Space Manifold: The space manifold $\mathscr{M}$ is a three-dimensional Euclidean space . $\mathbb{R}^3$.

- Time Manifold: The time manifold $\mathscr{T}$ is Newtonian absolute time, which is commonly used in inertial system analysis. A space-time manifold $\mathscr{T} \otimes \mathscr{M} = \mathbb{R} \otimes \mathbb{R}^3$ is called a Galilean Manifold.
- Characteristic Function: The characteristic function (functional) is defined as

$$\begin{aligned} f &= \pi_\mu \gamma^\mu, \\ &= \pi_i \gamma^i - \mathcal{H}(\pi, \gamma)t, \end{aligned}$$

where γ is a path defined on $\mathscr{M}$, and ππππ is a vector defined on the velocity bundle. The mass is defined as $m = ||\pi||/||\dot{\gamma}||$. The characteristic function ff can be considered as a functional of the path defined on $\Gamma \otimes \dot{\Gamma}$.

- Characteristic Forms: The characteristic 1- and 2-forms, respectively, are defined as

$$\begin{aligned} \omega &= \pi_\mu d\gamma^\mu, \\ &= \pi_i d\gamma^i - \mathcal{H}dt, \end{aligned}$$

and

$$\begin{aligned} \Omega &= d\pi_\mu \wedge d\gamma^\mu, \\ &= d\pi_i \wedge d\gamma^i - d\mathcal{H} \wedge dt^\mu. \end{aligned}$$

Here the characteristic functional is assumed to be invariant under affine transformation on the space manifold,4 MM. In this notation, the action and Lagrangian, respectively, are expressed as

$$\mathscr{I}(\gamma) = \int \mathcal{L}dt,$$

and

$$\begin{aligned} \mathcal{L}dt &= (\pi_i d\dot{\gamma}^i - \mathcal{H})dt, \\ &= \pi_\mu d\gamma^\mu. \end{aligned}$$

This transformation from the Hamiltonian to Lagrangian is known as a Legendre transformation and the independent variables of L are now (γ,γ˙). Here the flat Riemannian metric gμμ=(−,+,+,+) correctly induces a Legendre transformation from $\mathcal{H}$ to $\mathcal{L}$.

From the geometrical framework introduced in the previous subsection, we now construct a dynamical path system of a mass point evolving under the Hamiltonian formulation.

Theorem 3.1

(*Hamilton*) In a general dynamical space, the following two trajectories are equivalent:

1. Trajectory that gives the extremal of variation of the action $\delta\mathscr{I}(\gamma) = 0.$,
2. Trajectory that satisfies the characteristic equations:

$$\frac{d\gamma^i}{dt} = \frac{\partial\mathcal{H}}{\partial\pi_i}, \quad \frac{d\pi_i}{dt} = -\frac{\partial\mathcal{H}}{\partial\gamma_i}, \quad \frac{d\mathcal{H}}{dt} = \frac{\partial\mathcal{H}}{\partial t}.$$

Proof

1⇒2: Applying a variational operator to the action, we obtain

$$\begin{aligned} \delta\mathscr{I}(\gamma) &= \delta\int_{\gamma}\omega, \\ &= \int_{\delta\gamma}\omega = 0, \end{aligned}$$

implying that the integration of the characteristic 1-form is independent of δγδγ and satisfies Principle 1. Then, the Hamiltonian must satisfy the characteristic equations.

2⇒12⇒1: Applying the variational operator to the Lagrangian, we obtain

$$\begin{aligned} \delta\mathcal{L} &= \delta(\pi_i\dot{\gamma}^i) - \delta_{\mathcal{H}}, \\ &= \delta\pi_i\dot{\gamma}^i + \pi_i\delta\dot{\gamma}^i - \left(\frac{\partial\mathcal{H}}{\partial\gamma^i}\delta\gamma^i + \frac{\partial\mathcal{H}}{\partial\pi_i}\delta\pi_i\right), \\ &= \delta\pi_i\dot{\gamma}^i + \pi_i\delta\dot{\gamma}^i - \left(-\dot{\pi}_i\delta\gamma^i + \dot{\gamma}^i\delta\pi_i\right), \\ &= \delta\pi_i\dot{\gamma}^i + \pi_i\delta\dot{\gamma}^i - \left(\pi_i\delta\dot{\gamma}^i + \dot{\gamma}^i\delta\pi_i\right) = 0, \end{aligned}$$

,

where i=1,2,3. Here we use integration by parts and assume zero variations at both ends of the path. Steps 2 and 3 in the derivation are obtained by substituting the characteristic equations. □□

The variation of the action, on the other hand, can be written as $\delta\mathscr{I}(\gamma) = \int dt\delta\mathcal{L}$, which yields $\delta\mathscr{I}(\gamma) = 0 \Rightarrow \delta\mathcal{L} = 0$. This is analogous to the extremal of the Gibbs free energy in an isothermal reversible system at equilibrium. Let us consider this analogy in more detail. As pointed out, when introducing Eq. (11), the characteristic 1-form of the isothermal reversible transition is equivalent to the Gibbs free energy. In this analogy, the Hamiltonian corresponds to the entropy of the system and the macroscopic system configuration may be determined by the extremal point of the characteristic 1-form, which corresponds to the Lagrangian. The analogies between the Hamiltonian formalism of mass points and thermodynamics of ideal gases is summarized in Table 1. This analogy will be pursued further in the next subsection.

Table 1. Analogies between isothermal reversible thermodynamics and statistical dynamics of particle paths

	adiabatic free expansion	isothermal reversible transition	statistical dynamics of paths
Order parameter	entropy:S	temperature:T	time:t
Momentum vector	pressure:p	volume:V	momentum:π
Space coordinate	volume:V	pressure:p	position:$\gamma(t)$
Hamiltonian	temperature:T	entropy:S	Hamiltonian:$\mathcal{H}$
Lagrangian	internal energy: $-dU = p\,dV - TdS$	Gibbs free energy: $dG = Vdp - SdT$	Lagrangian: $\mathcal{L}dt = \pi d\gamma - \mathcal{H}dt$

Statistical mechanics of trajectories

The curvilinear path defined in the previous subsection is the trajectory of the particle. This trajectory is considered as the microscopic basis for constructing a statistical mechanical analog of the thermodynamic system.

Definition 3.6

(Lagrangian probabilistic space) The probabilistic space $\{\Gamma, \mathfrak{P}(\Gamma), p(\gamma)\}$ is imposed on the general dynamic path space (see Definition 3.5) as follows:

1. Whole event: The set of whole events is the curvilinear-path space.

2. σ-Algebra: The σσ-algebra is a power set of Γ, denoted $\mathfrak{P}(\Gamma)$.
3. Probability measure: Consider a map such that

$$p : \Gamma \to [0,1] \in \mathbb{R} : \gamma \mapsto p(\gamma), \quad \sum_{\gamma \in \Gamma} p(\gamma) = 1.$$

This functional describes the probability density to realize the path γγ. Please note that we consider only those paths whose start and end points are fixed at ξ_1 and ξ_2, respectively. The initial momentum vector must be chosen to realize a classical path. The probabilistic space $\{\Gamma, \mathfrak{P}(\Gamma), p(\gamma)\}$ is assumed to satisfy the probability axioms proposed by Kolmogorov [8]. If a set ΓΓ with infinite degrees of freedom belongs to ℵ1, then a set P(Γ) must belong to ℵ2. Since we cannot mathematically justify the probability measure defined on such an infinite set, we present a formal treatment only. The existence of the measure can be verified in limited cases, as will be discussed later.

Definition 3.7

(Entropy of Paths) The entropy of the paths is defined as

$$\mathcal{S} = -\sum_{\gamma \in \Gamma} p(\gamma) \log p(\gamma), \tag{12}$$

according to Shannon [13]. Here Γ, γ∈Γ and p(γ) are defined in the Definition 3.6.

Consistency between the above definition of entropy and Hamiltonian formalism will be discussed later. At this stage, we lack detailed knowledge of the dynamics that govern path behavior. However, it seems natural to configure paths by the following principle.

Principle 2

(Maximum entropy principle) Path configuration is determined to maximize the entropy of the paths.

According to above principle, a probability p(γ) to observe a path γγ can be given by following theorem.

Theorem 3.2

(*Canonical ensemble*) The probability p(γ) defined as following two requirement are equivalent:

1. The path whose entropy is maximized under the following constraints

$$\sum_{\gamma \in \Gamma} p(\gamma) = 1, \tag{13}$$

$$\sum_{\gamma \in \Gamma} p(\gamma)\mathscr{I}(\gamma) = I. \tag{14}$$

The first constraint is conservation of probability. The second stipulates that an action averaged over all possible paths II, called a *classical action*, must exist.

2. The path whose probability p(γ)p(γ) is given as

$$p(\gamma) = \frac{\exp\left(-\beta \mathscr{I}(\gamma)\right)}{Z(\beta)}, \tag{15}$$

$$Z(\beta) = \sum_{\gamma \in \Gamma} \exp\left(-\beta \mathscr{I}(\gamma)\right), \tag{16}$$

where β is a constant to eliminate a dimension in the argument of exponential function.

Here, the functional integration Z(β)Z(β) is regarded as a partition functional by analogy with equilibrium statistical mechanics. Then, particle trajectories (paths) form a canonical ensemble.

Proof

To maximize the entropy of the constrained paths, we introduce Lagrange multipliers αα and ββ such that

$$\phi(p, \alpha, \beta) = -S + \alpha\left(\sum_{\gamma \in \Gamma} p(\gamma) - 1\right) + \beta\left(\sum_{\gamma \in \Gamma} p(\gamma)\mathscr{I}(\gamma) - I\right).$$

The following conditions must be satisfied:

$$\frac{\partial\phi(p,\alpha,\beta)}{\partial\alpha} = \sum_{\gamma\in\Gamma} p(\gamma) - 1 = 0,$$

$$\frac{\partial\phi(p,\alpha,\beta)}{\partial\beta} = \sum_{\gamma\in\Gamma} p(\gamma)\mathscr{I}(\gamma) - I = 0,$$

$$\frac{\partial\phi(p,\alpha,\beta)}{\partial p} = \log p(\gamma) + 1 + \alpha + \beta\mathscr{I}(\gamma) = 0.$$

Here we used the functional derivative rules presented in the Appendix. Solving the above equations gives

$$p(\gamma) = \frac{\exp\left(-\beta\mathscr{I}(\gamma)\right)}{\sum_{\gamma\in\Gamma}\exp\left(-\beta\mathscr{I}(\gamma)\right)}.$$

To ensure that $\exp^{(-\beta\mathscr{I}(\gamma))}$ converges when $\mathscr{I}(\gamma)\to\infty$, β must be positive. □□

The above theorem implies that introducing the entropy of paths described by Eq. 12 yields the canonical ensemble of equilibrium statistical mechanics. Therefore, it appears that the classical trajectory of a mass point can be interpreted as the equilibrium state among all possible paths. The following theorem should then naturally hold:

Theorem 3.3

(*The most probable path*) The following two trajectories are equivalent:

1. Trajectory that gives the extremal of variation of the action $\delta\mathscr{I}(\gamma) = 0.$,

2. Trajectory that gives the maximum probability in Eq. (15).

Proof

Applying the variational operator δδ to both sides of Eq. (15), we obtain

$$\delta p(\gamma) = -\beta\frac{\exp\left(-\beta\mathscr{I}(\gamma)\right)}{Z}\delta\mathscr{I}(\gamma).$$

Thus, $\delta p(\gamma) = 0$ and $\delta \mathscr{I}(\gamma) = 0$ are equivalent. Since β, Z, and exp' $(-\beta \mathscr{I}(\gamma))$ are positive, the path γγthat minimizes I(γ)I(γ) gives the maximum p(γ). □□

Above theorem posits that the classical trajectory described in Theorem 3.1 is the most probable path of a mass point under Principle 2. Thus, we have proved that the maximum entropy principle (Principle 2) and Cartan principle (Principle 1) are mathematically equivalent in a general dynamic space.

4 CLASSICAL MECHANICS TO QUANTUM MECHANICS

This section relates the formulations of classical and quantum mechanics. Quantum and classical mechanics differ most distinctly by the probability amplitude, whose square gives probability density. Quantum mechanical motions, embodied in the Heisenberg/Schrödinger equation, are governed by the time evolution of probability amplitudes rather than probability densities. Importance of the quantum probability amplitude is discussed elsewhere [10]. The following subsection introduces a general framework of quantum mechanics, within which we relate our formulation to path integrals and stochastic quantization.

General quantum system

Here we develop a general framework for defining quantum mechanical probability amplitudes.

Definition 4.1

(Quantum amplitude 5 *and Probability)*

Let KK be any field, not necessarily commutative, and let VV be a linear (vector) space on it. The base field KKis associated with each element of ΓΓ. The probability amplitude and probability measure are introduced on these spaces as follows:

1. We introduce the following map from the path to an *n*-dimensional vector space:

$$\psi^n_{\mathbb{K}} : \Gamma \to V^n = \underbrace{V \otimes \cdots \otimes V}_{n} : \gamma \mapsto \psi^n_{\mathbb{K}}(\gamma),$$

where the paths have fixed end points at ξ_1 and ξ_2 (see Definition 3.1). The amplitude is then defined as

$$\psi^n_{\mathbb{K}}(\xi_2;\xi_1) = \sum_{\gamma\in\Gamma} \psi^n_{\mathbb{K}}(\gamma).$$

In defining a measure on the vector space, a σσ-algebra of the probability amplitudes in the base field K is assumed. Hereafter, the simplified notation ψKn denotes that the start and end points are fixed at ξ_1 and ξ_2, respectively, unless otherwise stated.

2. The following map from an amplitude to a real number

$$\mu : V^n \to \mathbb{R} : \psi^n_{\mathbb{K}} \to \mu\left(\psi^n_{\mathbb{K}}\right) \in [0,1],$$

is called a quantum probability (measure). The sequential map

$$\mu \circ \psi^n_{\mathbb{K}} : \Gamma \to (V^n \to)\mathbb{R} : \Gamma \mapsto \mu = \mu\left(\Gamma\right) = \mu(\psi^n_{\mathbb{K}}),$$

is also called a quantum probability and is represented by the same symbol μ(γ), where K and VV are as previously defined. The quantum probability measure must satisfy μ(ψKn)≤1. This measure is not normalized to unity because the curvilinear-path space Γ includes only paths with fixed end points; however, the quantum mechanical uncertainty relation precludes precise determination of an end point.

Two essential differences exist between the above-described general quantum system and canonical ensemble introduced earlier:

1. Relaxation of the first constraint in Theorem 3.2.1.
2. The probability amplitude is not necessarily a real number.

These differences may lead the dynamical system to adopt quantum mechanical instead of classical behavior.

Path integral quantization

Here we derive the probability measure and amplitude from the maximum entropy principle (Principle 2). The probability that a mass point observed at ξ_1 is later observed at ξ_2 is

$$p(\xi_2, \xi_1) = \mu(\Gamma) = \left| \sum_{\gamma \in \Gamma} \psi_{\mathbb{C}}^1(\gamma) \right|^2, \tag{17}$$

where the quantum probability amplitude $\psi_{\mathbb{C}}^1(\gamma)$ resides in a one-dimensional vector space on a complex number field C. Hereafter, $\psi_{\mathbb{C}}^1(\gamma)$ is written as $\psi(\gamma)$ for simplicity.

Theorem 4.1

(*Path integral quantization*) The quantum probability amplitude and probability measure that minimize the entropy

$$S = -\sum_{\gamma \in \Gamma} \left|\psi(\gamma)\right|^2 \log \left|\psi(\gamma)\right|^2, \tag{18}$$

under constraints

$$\sum_{\gamma \in \Gamma} \left|\psi(\gamma)\right|^2 \mathscr{I}(\gamma) = I, \tag{19}$$

$$\sum_{\gamma \in \Gamma} \left|\psi(\gamma)\right|^2 = 1, \tag{20}$$

is given by

$$\psi(\gamma) \simeq C\, e^{\frac{i}{\hbar}\mathscr{I}(\gamma)}, \tag{21}$$

where C∈RC∈R is an appropriate normalization constant and II is rendered dimensionless by dividing by the constant ℏℏ.

Proof

The probability amplitude that maximizes the entropy is again obtained by the Lagrange multiplier method:

$$\begin{aligned}\phi(\psi,\alpha,\beta) &= -S+\alpha\left(\sum_{\gamma\in\Gamma}\left|\psi(\gamma)\right|^2-1\right)+\beta\left(\sum_{\gamma\in\Gamma}\left|\psi(\gamma)\right|^2\mathscr{I}(\gamma)-I\right),\\ \frac{\partial\phi(\psi,\alpha,\beta)}{\partial\alpha} &= \sum_{\gamma\in\Gamma}\left|\psi(\gamma)\right|^2-1=0,\\ \frac{\partial\phi(\psi,\alpha,\beta)}{\partial\beta} &= \sum_{\gamma\in\Gamma}\left|\psi(\gamma)\right|^2\mathscr{I}(\gamma)-I=0,\\ \frac{\partial\phi(\psi,\alpha,\beta)}{\partial\psi^*(\gamma)} &= \psi(\gamma)\left(\log\left|\psi(\gamma)\right|^2+1+\alpha+\beta\mathscr{I}(\gamma)\right)=0.\end{aligned}$$

Here we again used rules of a functional derivative described in Appendix. Solving the above equations, the probability amplitude is obtained as

$$\begin{aligned}\psi(\gamma) &= e^{-\frac{1}{2}\beta\mathscr{I}(\gamma)}/Z^{1/2},\\ Z &= \sum_{\gamma\in\Gamma}e^{-\Re[\beta]\mathscr{I}(\gamma)}.\end{aligned}$$

Here $\beta\in\mathbb{C}$ and $\Re[\beta]$ $(\Im[\beta])$ are real-part (imaginary) of β, respectively. An imaginary part of β, is arbitrary due to an U(1) symmetry of $\psi(\gamma)$. As shown in the Theorem 3.3, a main contribution for the probability comes from the trajectory of the classical mass point. Moreover it is expected that a quantum mechanical path is small fluctuated around this classical trajectory. Thus here we assume that a real-part of ββcontributes to $\psi(\gamma)$. as slow moving function and replace it by a mean value. Then the probability amplitude can be express as

$$\begin{aligned}\psi(\gamma) &= \frac{1}{Z^{1/2}}e^{-\frac{1}{2}\beta\mathscr{I}(\gamma)},\\ &= \frac{1}{Z^{1/2}}e^{-\frac{1}{2}(\Re[\beta]+i\Im[\beta])\mathscr{I}(\gamma)},\\ &\simeq Ce^{-\frac{i}{\hbar}\mathscr{I}(\gamma)}.\end{aligned}$$

Equations (21) and (17) are nothing but the path-integral representation of transition probability introduced by Feynman [3, 4].6 The constant $\hbar\hbar$ is not necessarily the Planck's constant and cannot be determined within this formulation. Instead, a transition from classical to quantum mechanics arises through the probability amplitude, which can be a complex valued functional in our interpretation, in contrast to the real probability density of classical mechanics. This transition becomes evident if Eqs. (12) and (14) are compared with Eqs. (18) and (19).

CONCLUSIONS AND DISCUSSIONS

In this report, we introduced a general dynamic space that allows a unified geometric viewpoint of various dynamic systems. System dynamics were geometrically introduced though Cartan's principle. The equations of motion derived from Cartan's principle were found to be mathematically equivalent to Hamiltonian dynamics. Under the proposed generalized framework, the dynamics of a mass point were equivalent to those of equilibrium thermodynamics, enabling the derivation of a thermodynamic analogue of mass point dynamics. In fact, the maximum entropy principle defined in trajectory space generated precisely the Hamiltonian equation of motion. The classical trajectory of a mass point can be interpreted as the most probable path of the point. By extending the maximum entropy principle to probability amplitude rather than probability density, we retrieved the equations of path-integrated quantum mechanics. The probability amplitude was essential for transferring the system from a classical to quantum state. In summary, we incorporated various dynamical systems such as classical mechanics of a mass point and equilibrium thermodynamics and quantum mechanics of a point particle into a general mathematical framework.

While this framework provides a unique vantage point for both classical and quantum mechanics, it is not yet suitable for quantum mechanical analysis. We defined quantum amplitudes on a curvilinear space of precisely fixed end points. A basic quantum mechanical element is not naturally located in such a space-time manifold because of violation of the uncertainty relation. To satisfy the uncertainty relation, the essential element of quantum mechanics must be defined on a measurable space. A suitable candidate manifold is the Cartan tube introduced in Sect. 2.1, which is defined in a measurable extended phase space. The elements of this space, space-time and momentum manifolds comprise a Fourier-dual pair [9]. Thus, we can expect to construct quantum mechanics that satisfy the uncertainty condition on the Cartan tube. Moreover this formalism is suitable to treat quantum field theory, which is considered as a more fundamental theory than quantum mechanics of a mass point. A detailed analysis of this subject is beyond the scope of this report and will be reported elsewhere.

ACKNOWLEDGMENTS

We wish to thank Dr. Y. Sugiyama, Profs. T. Kon and J. Fujimoto for their continuous encouragement and fruitful discussions. The authors would like to thank Enago (www.enago.jp) for the English language review.

REFERENCES

1. Arnol'd, V.I., Vogtmann, K., Weinstein, A.: Mathematical methods of classical mechanics. Graduate Texts in Mathematics. Springer, Berlin (1989)
2. Cartan, E.: Leçons sur les invariants intégraux (1922)
3. Feynman, R.P., Brown, L.M.: Feynman's thesis: a new approach to quantum theory. World Scientific Publishing Company, Singapore (1942)
4. Feynman, R.P., Hibbs, A.R.: Quantum mechanics and path integrals. International Series in Pure and Applied Physics. McGraw-Hill, New York (1965)
5. Flanders, H.: Differential forms with applications to the physical sciences. Dover Books on Mathematics. Dover Publications, New York (1963)
6. Gemmer, M., Michel, J., Mahler, G.: Quantum thermodynamics : emergence of thermodynamic behavior within composite quantum systems, 2nd edn. Springer, Berlin (2009)
7. Kaku, M.: Introduction to superstrings and M-Theory. Graduate Texts in Contemporary Physics. Springer, New York (1999)
8. Kolmogorov, A.N.: Fundation of probability theory. Moscow-Leningrad (1936) (in Russian)
9. Kurihara, Y.: Classical information theoretic view of physical measurements and generalized uncertainty relations. J. Theor. Appl. Phys. **7**(1), 28 (2013)
10. Kurihara, Y., Quach, N.M.U.: Advantages of the probability amplitude over the probability density in quantum mechanics. ArXiv e-prints (2013)

11. Landau, L.D., Lifshitz, E.M., Pitaevskii, L.P.: Statistical physics: theory of the condensed state. Course of Theoretical Physics. Butterworth-Heinemann, UK (1980)

12. Lieb, E.H., Yngvason, J.: The physics and mathematics of the second law of thermodynamics. Phys. Rep. **310**(1), 1–96 (1999)

13. Shannon, C.E.: A mathematical theory of communication. Bell Syst. Tech. J. **27**(379–423), 623–656 (1948)

14. Story, T.L.: Dynamics on differential one-forms. iUniverse (2002)

15. Yamamoto, Y., Nakamura, K.: Analytical Mechanics 1. Asakura Publishing Co. Ltd., Tokyo (1998).

7

CHAPTER

Chromia Evaporation in Advanced Ultra-Supercritical Steam Boilers and Turbines

R. Holcomb Gordon[1]

[1] *National Energy Technology Laboratory, United States Department of Energy, Albany, OR, USA*

1. INTRODUCTION

Goals of the U.S. Department of Energy's Advanced Power Systems Initiatives include power generation from coal at 60% efficiency, which requires steam conditions of up to 760° C and 340 atm, so-called advanced ultra-supercritical (A-USC) steam conditions. Similar efforts are underway in Europe, Japan, and China to raise efficiencies by raising steam temperatures to 700° C and above. A limitation to achieving the goal is a lack of cost-effective metallic materials that can perform at these temperatures and pressures. Some of the more important performance limitations are high-temperature creep strength, fireside corrosion resistance, and steam-side oxidation resistance. Nickel-base superalloys are expected to be the materials best suited for steam boiler and turbine applications above about 675° C (Viswanathan et al., 2005). Specific alloys of interest include Haynes 230 and 282, Inconel 617, 625, and 740, Nimonic 263. Alloy compositions are given in Table 1 (Haynes International, 2008a, 2008b; Special Metals Corporation, 2004a, 2004b, 2005, 2006).

Steam-side oxidation can result in several adverse effects: general section loss from material thinning, deep localized section loss from internal oxidation (that may also provide crack initiation sites), reduced heat transfer due to the development of insulating oxide layers (that may lead to higher metal temperatures to maintain the same steam conditions), dimensional changes that are critical in airfoils, and downstream erosion from oxide spallation. Evaporation of protective chromia scales may also be an issue at the higher temperatures and pressures of A-USC steam turbines. The evaporation of chromia scales in steam is the focus of the research presented here.

Water vapor is known to adversely affect the oxidation behavior of chromia-forming alloys. Several explanations have been developed to explain the observed behaviors (Essuman et al., 2008; Young & Pint, 2006). Some are briefly described:

- Increased chromia growth rates from the doping of Cr_2O_3 with hydrogen species that result from the oxidation reaction between the metal and H_2O (Norby, 1993; Hänsel et al., 2003 as cited inYoung & Pint, 2006). Similar increased growth rates were found for the oxidation of Cr containing dissolved hydrogen compared to Cr without dissolved hydrogen (Henry et al., 2001 as cited in Essuman et al., 2008).
- Enhanced growth stresses that lead to scale cracking (Schütze et al., 2004 as cited in Essuman et al., 2008).
- Enhanced internal oxidation of Cr by water vapor changing the solubility or diffusivity of oxygen in the alloy (Essuman et al., 2008). Increases in internal oxidation can also lead to scales with less Cr_2O_3 formation and thus increased tendencies to breakaway oxidation (Essuman et al., 2008).
- Rapid, or "breakaway" oxidation from the depletion of Cr in the scale resulting from reactive evaporation of Cr_2O3 in the presence of O_2 and H_2O (Essuman et al., 2008; Holcomb, 2008,2009; Young & Pint, 2006). Clear links to evaporation were shown by how increased gas flow rates led to faster breakaway oxidation and increased Fe:Cr ratios within the oxide scale (Asteman et al., 2000).

This last behavior based on reactive evaporation of Cr_2O_3 is the subject of the work presented here. In prior work, a methodology to calculate the expected chromia evaporation rate as a function of temperature, pressure, gas velocity, and steam chemistry was developed for chromia forming alloys (Holcomb, 2008). The methodology was experimentally validated at low pressures and gas velocities by how well its kinetics predictions matched atmospheric pressure oxidation experiments in air plus water vapor environments. At 760° C and 340 atm, evaporation rates as high as 5.18×10^{-8} kg/m^2/s of $CrO_2(OH)_2(g)$ were predicted for rotating high pressure (HP) turbine blade components (Holcomb, 2008). This is equivalent to 0.08 mm per year of solid Cr loss. It was speculated that chromia evaporation upstream of the HP turbine, such as in the superheater, could partially saturate the steam with $CrO_2(OH)_2(g)$ and reduce the chromia evaporation rate.

TABLE 1. Nominal alloy compositions (wt%, except for B which is in ppm) and bulk alloy Cr concentrations (C_{Cr}, mol/m^3) for alloys of interest (Haynes International, 2008a, 2008b; Special Metals Corporation, 2004a, 2004b, 2005, 2006). Source values that were ranges are listed as the midpoint of the range. Source values that were maximums are listed as half the maximum. For economic reasons it is not unusual for actual chemical compositions of materials delivered for use in power plants to be near the lower end of the specification. Only Nb is listed for sources that gave a value or range for Nb+Ta. Source values for Pb, P, S and Bi are omitted.

Alloy	Fe	Cr	Ni	Co	Mo	Ti	Al	Mn	Other	C_{Cr}°
Haynes 230	1.5	22	Bal	2.5	2		0.3	0.5	0.02 La 14 W 0.4 Si 0.1 C 75 B	38000
Haynes 282	0.75	19.5	Bal	10	8.5	2.1	1.5		0.15 Cu 0.075 Si 0.06 C 50 B	31000
Inconel 617	1.5	22	Bal	12.5	9	0.3	1.15	0.5	0.25 Cu 0.5 Si 0.1 C 30 B	35400
Inconel 625	2.5	21.5	Bal	0.5	9	0.2	0.2	0.25	3.65 Nb 0.25 Si 0.05 C	34900
Inconel 740	0.7	25	Bal	20	0.5	1.8	0.9	0.3	2 Nb 0.5 Si 0.03 C	38700
Nimonic 263	0.35	20	Bal	20	5.85	2.15	0.3	0.3	0.1 Cu 0.2 Si 0.06 C 25 B	32200

The above efforts were expanded (Holcomb, 2009) to include

1. the use of new determinations of the thermodynamic properties of $CrO_2(OH)_2(g)$ (Opila et al., 2007),
2. expanding the methodology from flat plate components to cylindrical geometries (such as found within superheaters and steam pipes),

3. linking the chromia evaporation in steam to Cr diffusion within the alloy to predict alloy Cr concentration profiles and breakaway oxidation times, and

4. applying the breakaway oxidation predictions to a hypothetical superheater-steam pipe-HP turbine steam path, where the effects of $CrO_2(OH)_2(g)$ saturation along the steam path are applied. In many aspects this was similar to the work of Young and Pint (2006).

Presented here is further validation of the model by examining the reactive evaporation effects resulting from exposure of Haynes 230 and Haynes 282 to moist air environments as a function of flow rate and water content. These two alloys differ in Ti and Mn contents, which may form outer layers of TiO_2 or Cr-Mn spinels, which would in theory decrease the evaporation of Cr_2O_3 from the scale (Holcomb & Alman, 2006).

2. CHROMIA EVAPORATION

The oxidation of alloys protected by the formation of Cr_2O_3 (chromia formers) can undergo scale loss due to reactive evaporation of chromium-containing gas species. Water vapor increases the evaporation loss by allowing the formation of $CrO_2(OH)_2(g)$, which for these conditions has a higher vapor pressure than $CrO_3(g)$. $CrO_3(g)$ is the predominate Cr gas specie in dry air or oxygen (Graham & Davis, 1971). The formation of $CrO_2(OH)_2(g)$ can be written as:

$$\tfrac{1}{2}Cr_2O_3(s) + H_2O(g) + \tfrac{3}{4}O_2(g) = CrO_2(OH)_2(g) \tag{1}$$

The equilibrium partial pressure of $CrO_2(OH)_2(g)$ can be found from Eq. 2, where ΔG_1 is the Gibbs energy of reaction 1, R is the gas constant, T is the absolute temperature, K_1 is the equilibrium constant of reaction 1, a_{Cr2O3} is the activity of Cr_2O_3, and P_i is the partial pressure of gas specie i.

$$\Delta G_1^\circ = -RT\ln K_1 = -RT\ln\frac{P_{Cr2(OH)2}}{a_{Cr2O3}^{1/2}P_{H2O}P_{O2}^{3/4}} \tag{2}$$

Even while maintaining a Cr_2O_3 scale, evaporation can change the overall oxidation kinetics from parabolic behavior to linear kinetics or even to breakaway oxidation. Linear kinetics can arise after scale growth from oxidation, which decreases with increasing scale thickness, matches the scale loss from reactive evaporation (Tedman, 1966). The change in scale thickness, x, with time, t, can then be described in terms of the parabolic rate constant, k_p, and the linear reactive evaporation rate, k_e, as:

$$\frac{dx}{dt} = \frac{k_p}{x} - k_e \tag{3}$$

At long times or high reactive evaporation rates, a limiting scale thickness, x_L, arises that is given by:

$$x_L = \frac{k_p}{k_e} \tag{4}$$

In this case metal loss rates are linear, but still involve diffusion through a protective scale. Rapid metal loss can occur when reactive evaporation of Cr depletes the scale (and sometimes the substrate metal) of Cr (Asteman et al., 1999, 2000). Decreased Cr in the scale or metal can lead to the formation of less protective oxides, such as Fe-Cr oxides in Fe-Cr base alloys. Unprotective scales can lead to rapid metal loss, or "breakaway" oxidation.

A detailed methodology for calculating evaporation rates in a variety of environments was presented in earlier work for gas flow over a flat plate (Holcomb, 2008). Two basic equations were developed: Eq. 5 for laminar flow and Eq. 6 for turbulent flow:

$$k_e\left(\frac{kg}{m^2 s}\right) = 0.664\, Re_L^{0.5} Sc^{0.343} \frac{D_{AB} M_{CrO_2(OH)_2}}{LRT} P_{CrO_2(OH)_2} \tag{5}$$

$$k_e\left(\frac{kg}{m^2 s}\right) = 0.0592\, Re_L^{0.8} Sc^{0.333} \frac{D_{AB} M_{CrO_2(OH)_2}}{LRT} P_{CrO_2(OH)_2} \tag{6}$$

where Re_L and Sc are the dimensionless Reynolds and Schmidt numbers, D_{AB} is the gaseous diffusion coefficient between $CrO_2(OH)_2$ and the solvent gas (m^2/s), $M_{CrO2(OH)2}$ is the molecular mass of $CrO_2(OH)_2$ (kg/g-mol), L is the length (m) in the flow direction of the flat plate, $P_{CrO2(OH)2}$ is the partial pressure of $CrO_2(OH)_2$ (atm), R is the gas constant (8.20594×10^{-5} m^3atm/K g-mol), and T is the absolute temperature (K). The dimensionless Reynolds and Schmidt numbers are defined as:

$$Re_L = \frac{\rho_s u L}{\eta} \quad (7)$$

$$Sc = \frac{\eta}{\rho_s D_{AB}} \quad (8)$$

where ρ_s is the density of the solvent gas (kg/m^3), η is the absolute viscosity (kg/m/s), and u is the gas velocity (m/s). Values for two key parameters in Eqs. 5-8 (D_{AB} and η) are not established and require estimates (Holcomb, 2008) to be made.

The diffusion coefficient, D_{AB}, was estimated with a method using a $T^{1.75}/P_T$ relationship and "diffusion volumes" (Fuller et al., 1966; Holcomb, 2008), where P_T is total pressure. This estimate was later improved (Holcomb, 2009) using a computed geometry for $Cr_2(OH)_2(g)$ (Opila et al., 2007) to estimate the "diffusion volumes". The diffusion coefficient estimate in supercritical steam was further adjusted (reduced) based on methods presented by Bird et al. (1960) and Slattery and Bird (1958).

The absolute viscosity, η, was estimated for gas mixtures using the semi empirical formulation of Wilke (Wilke, 1950 as cited in Gaskell, 1992) that combines the absolute viscosity of the component gases. For non-polar gases, η was estimated based on the Lennard_jones potential (Gaskell, 1992). For water and steam, information of η as a function of T and P was used (Stultz & Kitto, 1992).

The Gibbs energy for reaction 1, (Opila et al., 2007), was used to determine $P_{CrO2(OH)2}$. This is a change from the prior work (Holcomb, 2008), where the thermodynamic data of Gindorf et al. (2001) was used. The differences in Gibbs energies for reaction 1 between the two transpiration studies (Gindorf et al., 2001, Opila et al., 2007) are shown in Table 2. The Opila values result in higher $P_{CrO2(OH)2}$ values (Eq. 2) and was attributed to possible undercounting of Cr loss in the Gindorf study and to the wider range in water vapor contents that allowed more precision in measurements at higher water levels (Opila et al., 2007).

The above methodology for flat surfaces (Eqs. 5-6) was expanded for use in pipes and to include possible saturation effects (Holcomb, 2009). Equations 5-6 were written in more general terms that included the average Sherwood number (Sh_{Ave}) and that allowed for the value of the partial pressure of $CrO_2(OH)_2(g)$ well away from the metal surface, $P_{CrO2(OH)2}$, to slow down the evaporation (Eqs. 4-5 assumed $P_{CrO2(OH)2}$ was zero), Eq. 9.

Here Sh_{Ave} is equal to $0.664Re_L^{0.5}Sc^{0.343}$ for laminar flow over flat plates and to $0.0592Re_L^{0.8}Sc^{0.333}$ for turbulent flow over flat plates.

TABLE 2. Simplified ΔG_T formulations based on the approximation that ΔG_T varies linearly with temperature for most substances (Kubaschewki & Alcock, 1979), where $\Delta G_T = \Delta H_x - T\Delta S_x$ and x is an elevated temperature close to the desired T.

Source	$\Delta H_x°$, J mol^{-1}	$\Delta S_x°$, J mol^{-1}	Notes
Opila et al., 2007	53500 ± 5400	-45.6 ± 6.6	Based on second law enthalpy and entropy methods from transpiration experiments to determine $\Delta H_{861}° = 53.5$ kJ mol^{-1} and $\Delta S_{861}° = 45.6$ J mol^{-1} K^{-1} at the average temperature of 861 K.
Gindorf et al., 2001	61400	-49.8	Based on transpiration experiments to determine $\ln P_{Cr2(OH)2}$ (Pa) to be equal to 0.442 + -7387.5T^{-1} with an average temperature of 948 K in air with 2% H_2O.

$$k_e\left(\frac{kg}{m^2 s}\right) = Sh_{Ave}\frac{D_{AB}M_{CrO_2(OH)_2}}{LRT}\left(P_{CrO_2(OH)_2} - P^{\circ}_{CrO_2(OH)_2}\right) \quad (9)$$

For flow within circular tubes, Eq. 9 was used but with a different expression for Sh_{Ave}. For the analysis that follows, the Dittus-Boelter equation (Winterton, 1998; Incropera & DeWitt, 2001) was used for Sh_{Ave} for turbulent conditions:

$$Sh_{Ave} = 0.023Re_d^{0.8}Sc^{0.4} \quad (10)$$

where Re_d is the same as Re_L but with the diameter, d, instead of the plate length, L, in Eq. 7. For rough pipes the use of an expression for Sh_{Ave} that incorporates a friction factor, such as that of Petukov (Incropera & DeWitt, 2001; Petukhov, 1970) can be used.

One additional point is that the evaporation rate, k_e, from Eq. 9 is for the amount of $Cr_2(OH)_2(g)$ leaving the surface, while the mass change data will be in terms of $Cr_2O_3(s)$ lost from the surface. The conversion is shown in Eq. 11 (Holcomb, 2008). Unless otherwise stated, k_e is on a $Cr_2(OH)_2(g)$ basis.

$$k_{e,\ Cr2O3} = \frac{M_{Cr2O3}}{2M_{CrO2(OH)2}} k_{e,\ CrO2(OH)2} = 0.644 k_{e,\ CrO2(OH)2} \qquad (11)$$

3. CHROMIA EVAPORATION IN CYCLIC OXIDATION TESTS

Cyclic oxidation tests were conducted for up to 2000 hourly cycles in moist air. Each hourly cycle consisted of 55 minutes in the furnace, withdrawal from the furnace, holding for 3.4 minutes, and then placement back into the furnace. The remaining 1.6 minutes per hourly cycle was the travel time the samples took into and out of the furnace. The samples experienced rapid heating and cooling rates, and were oriented such that the gas flow was parallel to their surfaces. More experimental details can be found in prior work (Holcomb et al., 2007).

A goal of the cyclic tests was to evaluate alloys for use in A-USC boilers and turbines. Scale exfoliation, resulting in part from thermal expansion mismatch between the alloy and the scale, is an important failure mechanism in boiler tubes (Wright & Tortorelli, 2007). The use of thermal cycles is an aggressive test of scale exfoliation tendencies. While some alloys did exhibit scale exfoliation in the cyclic tests (Holcomb et al., 2007), except as noted no scale exfoliation was observed during tests on the alloys discussed below.

Results for two alloys from prior work (Holcomb, 2008, 2009), Haynes 230 and Inconel 740, are shown in Figs. 1 and 2 respectively, and also in Table 3. The Haynes 230 results show the influence of gas velocity—a factor of 4 increase in gas velocity resulted in a factor of 2 increase in evaporation, which is consistent with Eqs. 7 and 9. The dashed lines in Fig. 1 show the evaporation rate from the model, with an activity of chromia of 0.05. This low chromia activity presumably arises from chromia depletion in the outer portion of the oxide scale as chromia is lost to evaporation. The results for two different heats of Inconel 740 in Fig. 2 show two different behaviors. The heat with the □ symbols initially had a similar evaporation rate as the Haynes 230 results. The evaporation rate decreased after about 800 cycles. The other heat, with + symbols, exhibited much less chromia evaporation. One difference between the two heats was 1.6 Ti in the heat with the □ symbols compared with 1.8 Ti in the heat with + symbols.

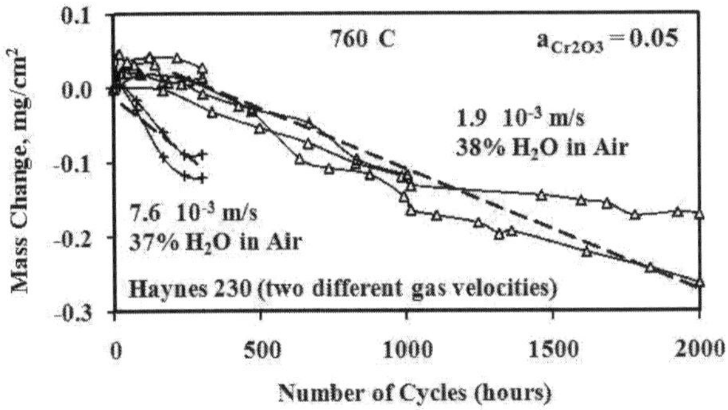

FIGURE 1. Mass change measurements for Haynes 230 exposed at 760° C in moist air at two different gas velocities. The dotted lines are the predicted mass change due to chromia evaporation with a chromia activity of 0.05. Adapted from Holcomb (2009).

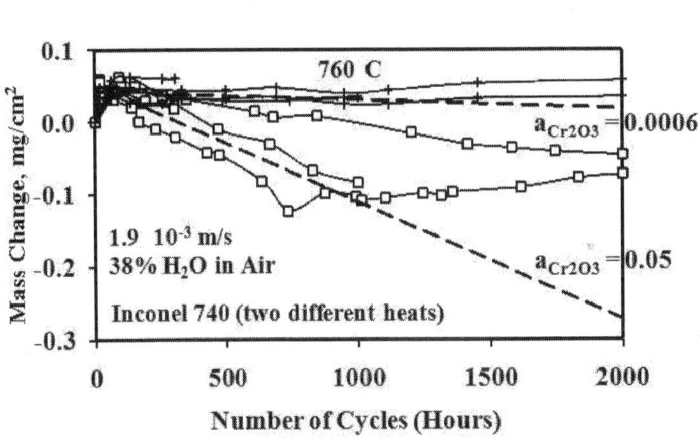

FIGURE 2. Mass change measurements for two different heats of Inconel 740 exposed at 760° C in moist air. The dotted lines are the predicted mass changes due to chromia evaporation with chromia activities of 0.05 and 0.0006. Adapted from Holcomb (2009).

Results for Haynes 230 and Haynes 282 at 800° C are shown in Figs. 3-4 for two water vapor levels and two gas flow rates, and are also shown in Table 3. The results for Haynes 230 continue to show the same gas velocity dependence shown at 760° C—a four-fold increase in flow rate doubles the evaporation rate. For Haynes 282, the combination of the highest flow rate and water vapor percentage changed behavior after 1000-12000 cycles. Past this point, small amounts of debris were found upon handling the sample for weighing. So the mass loss was a combination of chromia evaporation and scale spallation. A four-fold increase in flow rate increased, but did not double evaporation rate for Haynes 282. It could well be the case that steady state evaporation was not yet reached for the Haynes 282 prior to it starting to spall.

When the results in Figs. 3-4 were compared with the chromia evaporation model, chromia activities were found that matched the results—found in the last column in Table 3. One thought from prior research (Holcomb, 2009) was that chromia activities would decrease with higher rates of evaporation. As chromia leaves the surface, Mn and Ti oxides would become enriched and thus decrease the chromia activity. Table 3 shows that this is not always the case. Increasing the flow rate while maintaining a constant P_{H2O} resulted in chromia activities of Haynes 230 and Haynes 282 dropping from 0.14 to 0.11 and 0.21 to 0.07, respectively. However, raising the P_{H2O} from 0.19 to 0.57 while maintaining a constant flow rate increased the evaporation rate but also increased the chromia activity for Haynes 230 (0.06 to 0.11). The expected result was found for Haynes 282 with a decrease from 0.12 to 0.07.

Scanning electron microscopy (SEM) was used to examine the surfaces of the 800° C samples, Fig. 5. The surface structure of Haynes 230 was finer than that found on Haynes 282. The Haynes 282 at the fastest chromia evaporation rate (lower right-hand image in Fig. 5) shows holes in the structure. One could speculate that these holes were where chromia may have evolved from and thus undercut the outer oxide. This undercutting appears to have resulted in the scale spalling in this sample.

The use of x-ray diffraction (XRD) on these surfaces showed chromia and spinels on Haynes 203 and chromia, spinels, and TiO_2 on Haynes 282. Various Cr-Ni-Mn containing spinels have very similar XRD peaks, making identification of Mn-containing spinels impossible.

Elemental spot analyses on the Haynes 230 surface for the slowest chromia evaporation sample (19% H_2O, 4.2×10^{-3} m/s) did not show any Mn on the surface. Analyses of the other two samples, with higher evaporation rates, showed Mn throughout the surface (along with Cr and O).

Elemental spot analyses on all of the Haynes 282 surfaces showed two distinct types of areas within a fine structure. Figure 6 shows a higher magnification image of this surface and these two types of areas. The smaller grains are primarily Cr and O (with smaller amounts of Ni and Al), and the larger grains (with flat surfaces) to be essentially pure Ti and O.

TABLE 3. Alloys, conditions, and results for cyclic oxidation tests. Gas flow rates are at temperature. Data for 760° C are from prior work (Holcomb, 2008). Data for Inconel 740 are from the heat with □ points in Fig. 2.

Alloy	T, °C	P_{H2O}	Flow Rate, m/s	k_e [Cr_2O_3], kg/m²s	Time Range, hr	Apparent a_{cr2O3}
Haynes 230	760	0.37	1.9×10^{-3}	3.8×10^{-10}	200-2000	~0.05
Haynes 230	760	0.37	7.6×10^{-3}	11.7×10^{-10}	94-2000	~0.05
Inconel 740	760	0.38	1.9×10^{-3}	1.5×10^{-10} 5.5×10^{-10}	102-2000 18-734	~0.0006 ~0.05
Haynes 230	800	0.19	4.2×10^{-3}	3.4×10^{-10}	200-2000	0.06
Haynes 230	800	0.57	1.0×10^{-3}	4.9×10^{-10}	24-2000	0.14
Haynes 230	800	0.57	4.2×10^{-3}	9.1×10^{-10}	24-2000	0.11
Haynes 282	800	0.19	4.2×10^{-3}	4.9×10^{-10}	600-2000	0.12
Haynes 282	800	0.57	1.0×10^{-3}	6.1×10^{-10}	500-2000	0.21
Haynes 282	800	0.57	4.2×10^{-3}	7.2×10^{-10}	125-1000	0.07

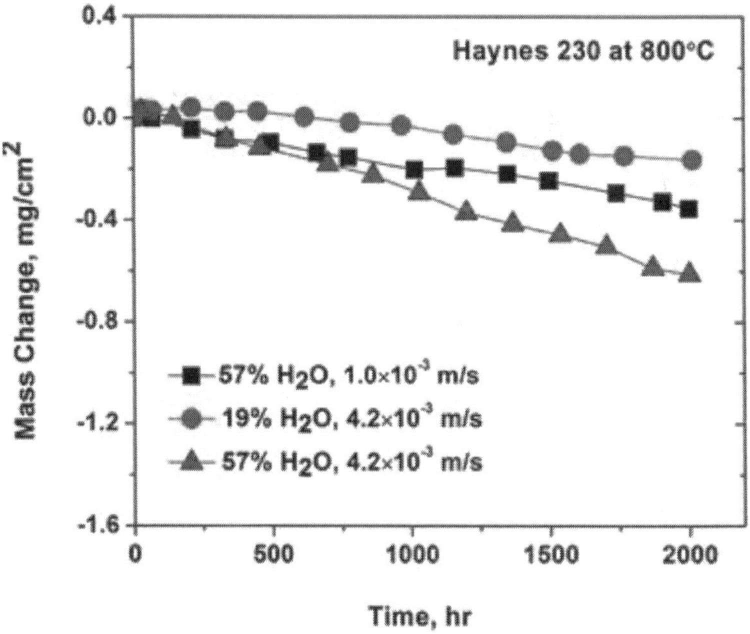

FIGURE 3. Mass change measurements Haynes 230 at 800° C in moist air at three different water vapor levels and gas flow rates.

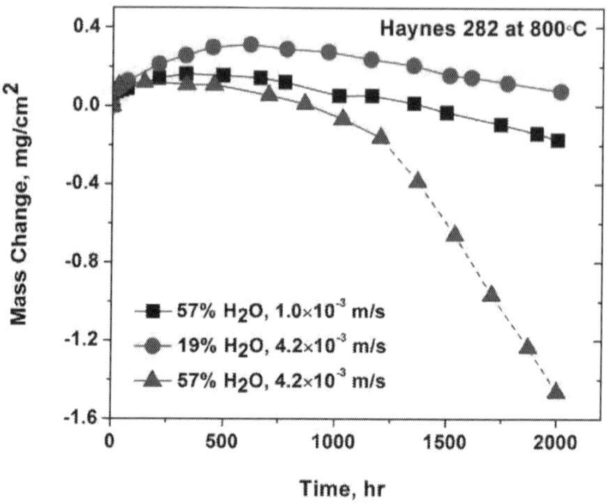

FIGURE 4. Mass change measurements Haynes 282 at 800° C in moist air at three different water vapor levels and gas flow rates. The dashed line indicates spalling.

4. CHROMIUM DIFFUSION

After evaporation causes the overall oxidation kinetics to become linear and to have a steady state scale thickness (Eq. 4), the flux of Cr away from the scale via evaporation must equal the flux of Cr to the scale via Cr diffusion within the alloy. If there is insufficient arrival of Cr to the surface, then the chromia scale will not be maintained and breakaway oxidation would be expected to eventually occur.

To model Cr diffusion within the alloy with a constant flux of Cr leaving the surface, the transient heat transfer model of Incropera and DeWitt (Incropera & DeWitt, 2001) within a semi-infinite solid and with a constant surface heat flux was used. The underlying mathematics between heat and mass transfer are the same, so this model was used with appropriate mass transfer parameters. The overall equation for the concentration of Cr as a function of x (depth from the surface into the alloy, m) and t (time, s), $C_{Cr}(x,t)$ is:

$$C_{Cr}(x,t) = C_{Cr}^{\circ} - \frac{2k_e}{M_{CrO_2(OH)_2}}\sqrt{\frac{t}{\pi D_{Cr}}}\exp\left(-\frac{x^2}{4D_{Cr}t}\right) + \frac{k_e x}{M_{CrO_2(OH)_2}D_{Cr}}\operatorname{erfc}\left(\frac{x}{2\sqrt{D_{Cr}t}}\right) \tag{12}$$

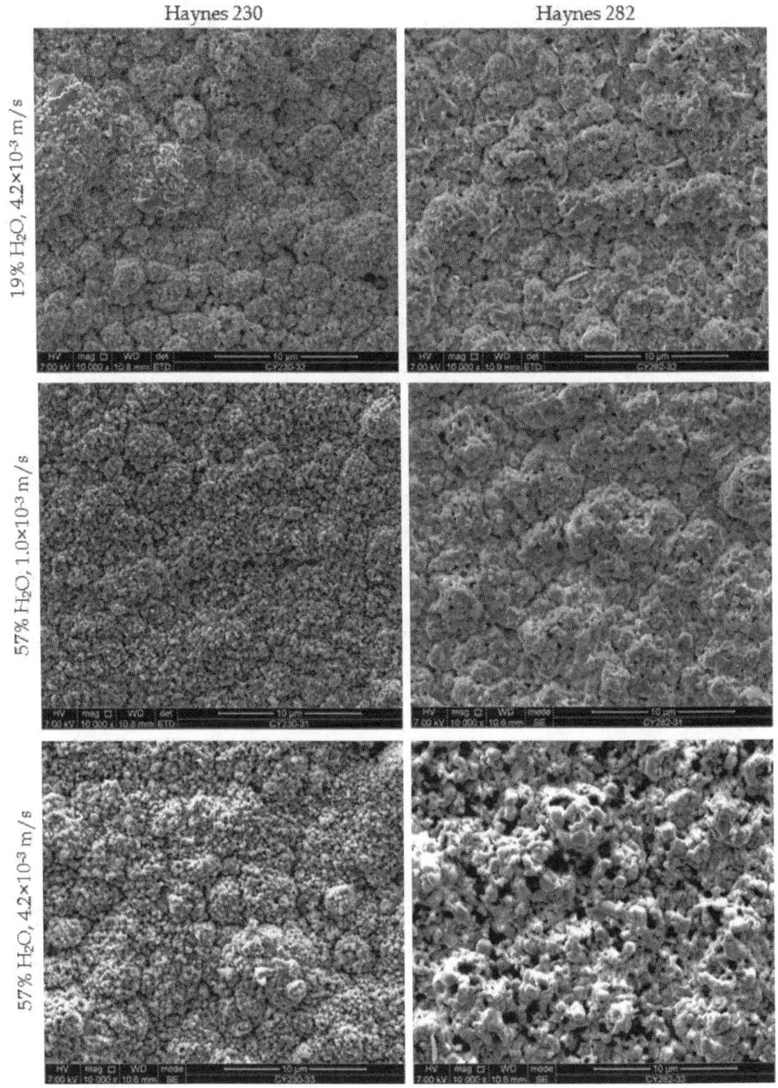

FIGURE 5. Surface SEM after 2000 cycles at 800° C in moist air.

FIGURE 6. Secondary electron image of Haynes 282 after 2000 hourly cycles at 800° C in moist air (57% H_2O) and a gas flow rate of 4.2×10^{-3} m/s. Elemental spot analysis shows the smaller grains to be primarily Cr and O (with smaller amounts of Ni and Al), and the larger grains (with flat surfaces) to be essentially pure Ti and O

where C_{Cr} is the bulk alloy Cr concentration (mol/m^3) (Table 1), D_{Cr} is the diffusion coefficient of Cr (m^2/s), and erfc is the complementary error function. For the concentration of Cr at the surface as a function of time, Eq. 12 simplifies to Eq. 13.

$$C_{Cr}(0,t) = C_{Cr}^{\circ} - \frac{2k_e}{M_{CrO_2(OH)_2}}\sqrt{\frac{t}{\pi D_{Cr}}} \tag{13}$$

And the surface concentration is equal to zero at time t*, as shown in Eq. 14.

$$t^* = \frac{\pi D_{Cr}}{4}\left(\frac{M_{CrO_2(OH)_2} C_{Cr}^{\circ}}{k_e}\right)^2 \tag{14}$$

Breakaway oxidation is expected to occur prior to t*, such as when the surface concentration becomes less than the critical value required to sustain a Cr_2O_3 scale. The critical value could be estimated, as was done for the critical bulk concentration of Cr in Fe-Cr alloys (Meier et al., 2010), if sufficient diffusion rates in the metal and scale are known. However, the lower bound on the critical value of the surface concentration is zero. So t* is used as a proxy for the time until breakaway oxidation will occur.

At the temperatures in a steam turbine, the diffusion coefficient, D_{Cr}, is an effective diffusion coefficient and is a combination of lattice, $D_{Cr}{}^{L}$, and grain boundary, $D_{Cr}{}^{gb}$, diffusion. The relationship used is based on cubic grains of size λ (m) and with grain boundary width δ (m) (Peng et al., 2005):

$$D_{Cr} \cong D_{Cr}^{L} + \frac{2\delta}{\lambda} D_{Cr}^{gb} \tag{15}$$

The lattice diffusion coefficients were determined using Dictra diffusion simulation software (Thermo-Calc Software AB, 2006) with the Ni-DATA (Saunders, 2000) and MOB2 (Royal Institute of Technology, 1999) databases. Figure 7 shows lattice diffusion coefficients of Cr in the face-centered-cubic (FCC) phase for several alloys of interest as a function of Cr content at 760° C. The right-most-point for each alloy is at the alloy Cr content. The lower Cr content values represent the alloy where it is locally depleted in Cr near the surface. Figure 7 shows that for many of the alloys an assumption of a constant value for the diffusion coefficient is a good one.

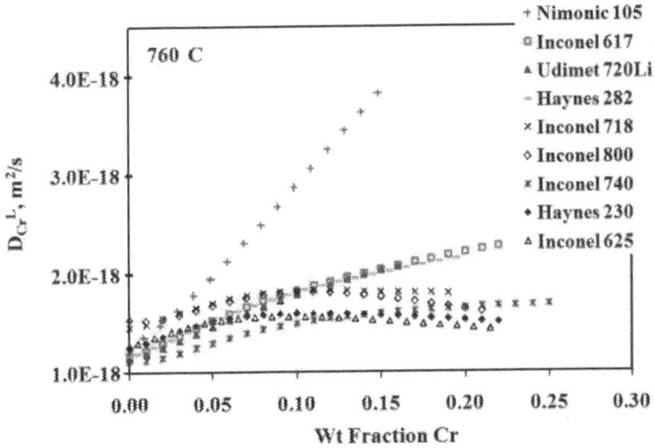

FIGURE 7. Lattice diffusion coefficients at 760° C for Cr in the FCC phase of nickel base superalloys determined using Dictra diffusion simulation software (Thermo-Calc Software AB, 2006) with the Ni-DATA (Saunders, 2000) and MOB2 (Royal Institute of Technology, 1999) databases.

The use of the FCC phase as the basis of lattice diffusion is a simplification of the rather complex nature of these alloys. All of the alloys in Fig. 7 are strengthened by the formation of one or more second phases, the amount of which depends upon heat treatment. Two of these alloys (Nimonic 105 and Udimet 720Li) have very significant amounts of a second phase (gamma prime). The ordered structure of these second phases should result an diffusivities that are much smaller than the FCC phase. So this simplification would tend to overstate the overall effective diffusion coefficients.

Further phase complications can arise from Cr depletion, where the loss of Cr can itself lead to phase changes. An example of this can be seen in Fig. 8, for Inconel 617 exposed to 2000 hourly cycles at 760° C in air with 37% H_2O at atmospheric pressure (Holcomb, 2009). Figure 8 shows a phase transition zone near the oxide interface, presumably from Cr depletion. Internal oxidation, mainly of aluminum, is also present underneath the oxide scale along grain boundaries. Thus, a more comprehensive phase-based treatment of diffusion coefficients may be warranted. However, for the calculations that follow, the values used for the lattice diffusion coefficients are ones from the FCC phase at the Cr content of the alloy (Fig. 7).

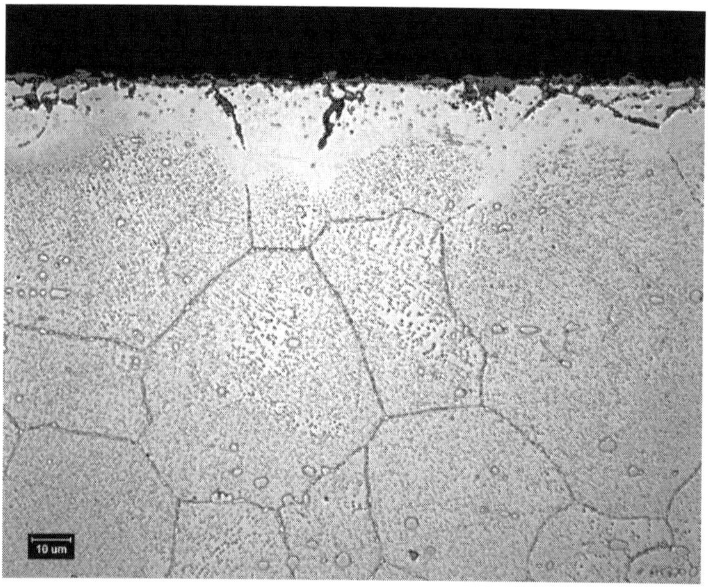

FIGURE 8. Results of cyclic oxidation of Inconel 617 at 760° C in air plus 37% H_2O after 2000 hourly cycles (Holcomb, 2009). A bright field image after etching (50% concentrated nitric acid, 50% glacial acetic acid). A precipitate free zone lies beneath internal oxidation areas.

Grain boundary diffusion coefficients are estimated based on the work of Paul *et al.* (Paul et al., 1994) where lattice and grain boundary diffusion coefficients were determined for Inconel 800 (where a grain boundary width of 0.5 nm was used). The estimate is to use the same ratio of grain boundary diffusion to lattice diffusion that was found for Inconel 800, and to use the same grain boundary width of 0.5 nm. This ratio is given by Eq. 16.

$$\frac{D_{Cr}^{gb}}{D_{Cr}^{L}} \cong \exp\left(\frac{12412}{T} - 1.7203\right) \tag{16}$$

An example of a comparison between calculated and measured Cr diffusion profiles is given in Figs. 9-10 for Inconel 740 exposed to 2000 hourly cycles at 760° C (under the same conditions as in Fig 8).Figure 9 is a backscattered electron image of a cross-section of the metal-scale interface. It shows a very thin oxide scale and a larger thickness of internal oxidation. The internal oxidation is primarily of aluminum and titanium. This kinetic data for this specimen are the points in Fig. 2 with the most mass loss. Figure 10 shows the Cr concentration profile measured using X-ray photoelectron spectroscopy (points) and the results of the model of Eqs. 12 and 16 with λ equal to 50 μm, δ equal to 0.5 nm (line), and k_e of 8.0×10^{-10} kg/m²s. This evaporation rate is the model prediction with an a_{Cr2O3} of 0.4. The model is a close fit to the measured points. Note that the use of 2000 hours in Fig. 10 is an approximation that overstates the time at temperature (each hourly cycle had a few minutes at lower temperatures). If the predicted evaporation rate using pure chromium is used (a_{Cr2O3} of 1), the predicted evaporation rate is 2.0×10^{-9} kg/m²s, and the model would predict too much Cr loss (a wt fraction of Cr of 0.05 at the surface).

The model curve in Fig 10 invoked an a_{Cr2O3} of 0.4. Is lowering the chromia activity valid? These alloys are not pure Ni-Cr alloys and have other alloying elements that oxidize, are mobile, and many times can be found on the outer surface of oxide scales—namely Ti and

Mn. For example, enough Mn additions to result in outer scales of Mn-Cr spinels instead of chromia is expected to reduce the evaporation of Cr by a factor of 55 at 700° C and by a factor of 35 at 800° C (Holcomb & Alman, 2006). This is equivalent to a chromia activity of 0.0006 at 760° C. Prior work (Holcomb, 2009) showed that the higher evaporation rates shown by Haynes 230 and Inconel 740 corresponded to an a_{Cr2O3} value of 0.05. The use of 0.05 in the analysis used for Fig. 10 results in a predicted evaporation rate of 2.8×10^{-10} kg/m²s, and a wt fraction of Cr of 0.22 at the surface. Since

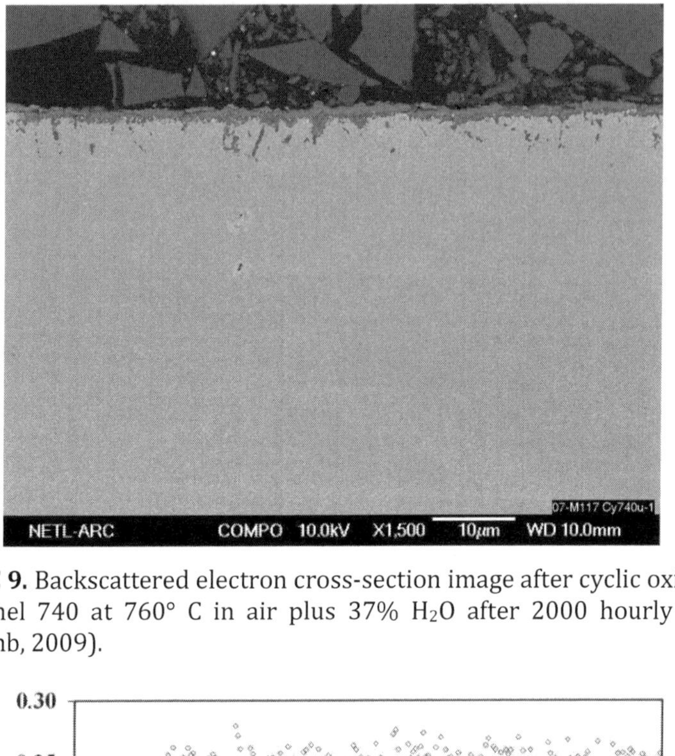

FIGURE 9. Backscattered electron cross-section image after cyclic oxidation of Inconel 740 at 760° C in air plus 37% H_2O after 2000 hourly cycles (Holcomb, 2009).

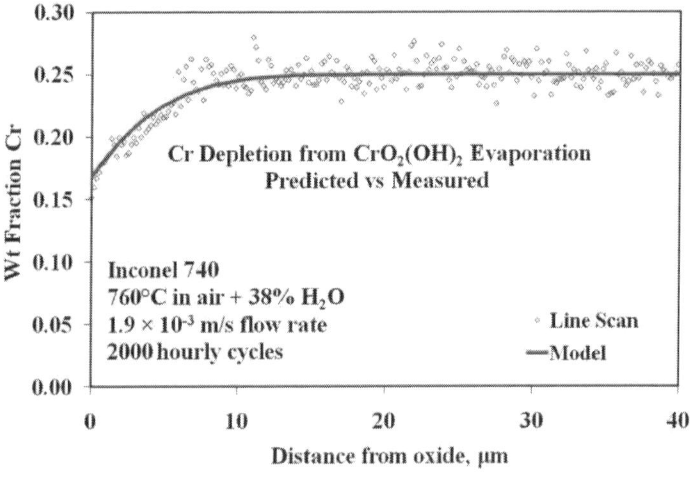

FIGURE 10. Chromium depletion from cyclic oxidation of Inconel 740 at 760° C in air plus 37% H_2O after 2000 hourly cycles. The individual points are from a line scan into the sample on the cross-section shown in Fig. 9 (Holcomb, 2009). The line results from the Eq. 12. The evaporation rate used in Eq. 12was 8×10^{-10} kg/m^{2}s, which corresponds to an a_{Cr2O3} of 0.4.

some of the Cr loss can be attributed to the non-evaporated chromia scale, the best value for a_{Cr2O3} lies between 0.05 and 0.4. Apparent a_{Cr2O3} values, based on the negative slope of the mass change data, are shown in Table 3 for each alloy and condition in this study. Most of the apparent a_{Cr2O3} values are within this range

5. HYPOTHETICAL SUPERHEATER-STEAM PIPE-HP TURBINE STEAM PATH

The combination of gas evaporation of $CrO_2(OH)_2$, gas saturation of $CrO_2(OH)_2$, and Cr depletion in the alloy is illustrated in Fig. 11. Figure 11 represents one section of superheater (SH) tubing or steam pipe of cell length L. The bulk partial pressure of $CrO_2(OH)_2$ coming into the cell is $P_{CrO2(OH)2}$(Incoming). Evaporation within the cell at rate k_e raises the $P_{CrO2(OH)2}$ (Outgoing) by the amount released by evaporation and is shown in Eq. 17.

$$P^{\circ}_{CrO_2(OH)_2\,(Outgoing)} = P^{\circ}_{CrO_2(OH)_2\,(Incoming)} + \frac{4k_e LRT}{duM_{CrO_2(OH)_2}} \tag{17}$$

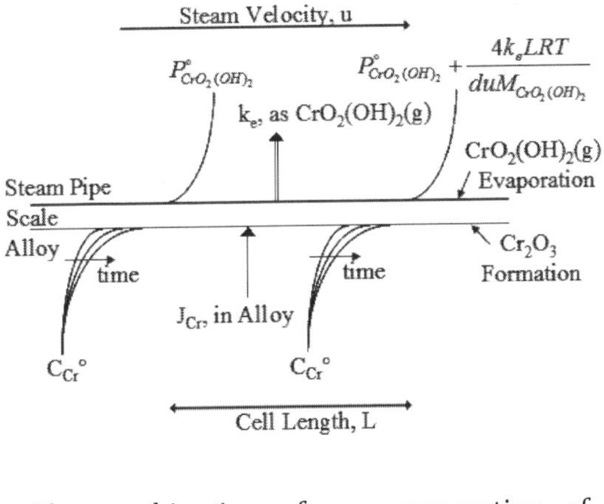

FIGURE 11. The combination of gas evaporation of $CrO_2(OH)_2$, gas saturation of $CrO_2(OH)_2$, and Cr depletion in the alloy is illustrated. This represents one section of superheater tubing or steam pipe of cell length L (Holcomb, 2009).

The effects of saturating the gas phase with $CrO_2(OH)_2$ can then be calculated along the length of a SH tube or steam pipe by combining many cells together. A hypothetical arrangement of a 100 m long, 0.05 m I.D., SH that leads into a 50m long, 0.3 m I.D., steam pipe that leads to a high pressure (HP) turbine is illustrated in Fig. 12. The temperature profile and steam velocities are shown on the left-hand axis, while the evaporation rate is shown on the right-hand axis. The evaporation rate is based on an a_{Cr2O3} of 0.05. The SH and steam pipe each consisted of 160 cells of the type shown inFig. 11. Saturation results in an 80% reduction in the evaporation rate at the HP turbine.The steam chemistry used in Fig. 12 and in subsequent calculations was based off of feedwater treatments that are typical of once-through supercritical power plants, i.e., a pH of 8.0-8.5 controlled with ammonia additions (Babcock & Wilcox, 1992). At high temperatures, water undergoes dissociation to O_2 and $H_2$to levels well above 1 ppb. To estimate the dissolved oxygen required for Eq. 1 at temperature and pressure, the program FactSage 5.5 (Bale et al., 2007) was used to first determine the amount of NH_3required for a pH of 8.25 at 25° C: 34.5 ppb. This agreed well with the reported (Babcock & Wilcox, 1992) 20-65 ppb NH_3 used for pH control to 8.0 to 8.5. Next FactSage 5.5 was used to find the value of the fugacities of H_2O and O_2 for each temperature and pressure combination from water with 34.5 ppb NH_3. The use of fugacities instead of partial pressures made only a minor difference because the fugacity adjustments tended to cancel each other out.

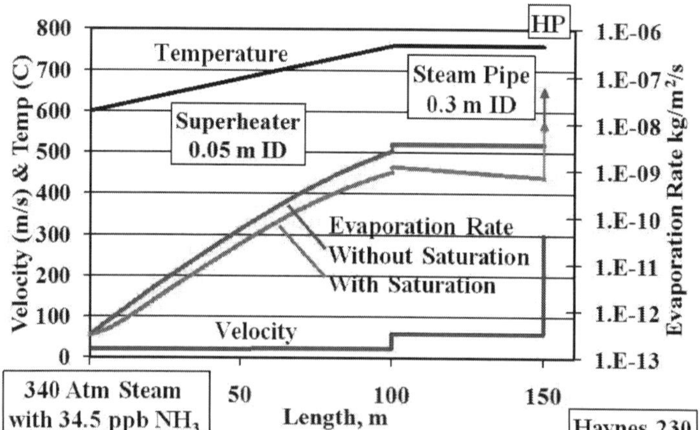

FIGURE 12. The effects of gas saturation is shown for a hypothetical arrangement of a 100m long SH that leads into a 50m steam pipe that leads to a high pressure (HP) turbine. Saturation results in a 80% reduction in the evaporation rate at the HP turbine. An a_{Cr2O3} of 0.05 was used. Adapted fromHolcomb (2009).

The time for the alloy Cr level to reach zero at the scale-metal interface, using Eq. 14, is shown in Fig. 13 for the same conditions as in Fig. 12 (and using saturated values for k_e). Haynes 230 was the alloy used. Under this simulation the chromia evaporation at the HP turbine blade is expected to be so great as to deplete the alloy surface of Cr in a few days. However, the very high rates of evaporation predicted at the HP turbine will likely be less than the model. The model is highly dependent upon the chromia activity term, especially in combination with the saturation effect. Should the higher evaporation rate at the HP lead to a lower chromia activity (while keeping the a_{Cr2O3} in the SH and steam pipe at 0.05), then the equilibrium surface $P_{Cr2(OH)2}$ can easily drop below $P_{Cr2(OH)2}$, and thus stop chromia evaporation (Eq. 9). In this example (with the a_{Cr2O3} in the SH and steam pipe at 0.05), an a_{Cr2O3} in the HP turbine section of 0.032 would be sufficiently low enough to stop chromia evaporation in the HP turbine.

The very high temperatures and total pressures of Figs. 12-13 match that of the U. S. Department of Energy's goals of 760° C and 340 atm. The impact of evaporation at lower temperatures and pressures can be seen in Figs. 14-15, where similar procedures are used to estimate the time it takes for the surface concentration of Cr in the alloy to reach zero. In all cases the SH inlet temperature was taken to be 600° C. Below 600° C, ferritic alloys, which would not be expected to form chromia scales, would typically be used.

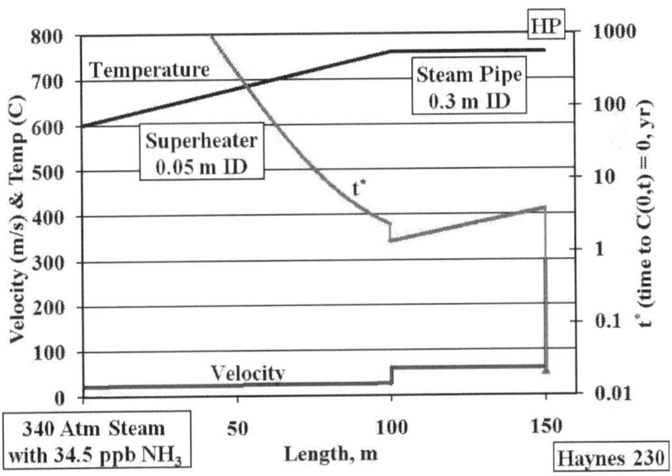

FIGURE 13. The predicted time for the Cr level at the scale-metal interface to reach zero. Haynes 230 was the alloy system used in this estimate. An a_{Cr2O3} of 0.05 was used. Adapted from Holcomb (2009).

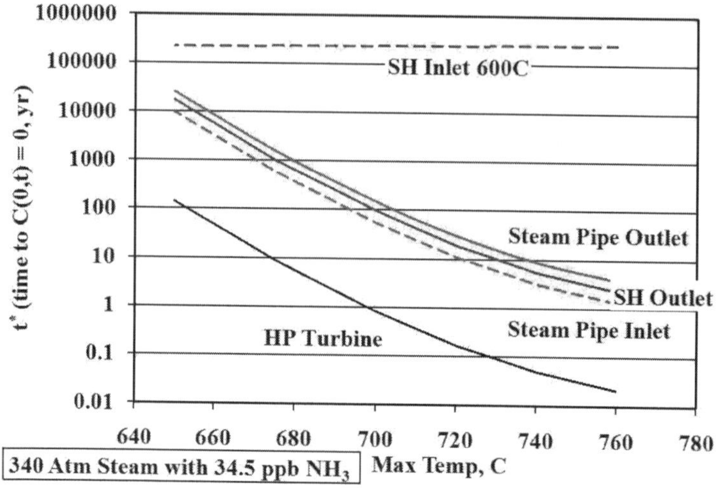

FIGURE 14. The effects of temperature on the estimate of the time it takes for the surface concentration of Cr in the alloy to reach zero, at 340 atm. Haynes 230 was the alloy system used in this estimate. An a_{Cr203} of 0.05 was used. Adapted from Holcomb (2009).

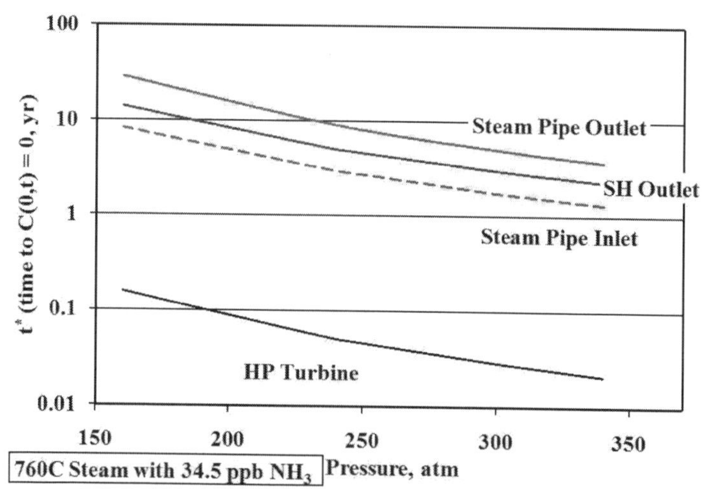

FIGURE 15. The effects of pressure on the estimate of the time it takes for the surface concentration of Cr in the alloy to reach zero, at 760° C. Haynes 230 was the alloy system used in this estimate. An a_{Cr203} of 0.05 was used. Adapted from Holcomb (2009).

Figure 14 shows why chromia evaporation is not considered to be of much concern in existing steam boilers and turbines. Modern advanced steam cycles are currently no higher than 630° C, so the time to breakaway oxidation (for which the time for C(0,t) to reach 0 is being used as an estimate) is well beyond the life span of the boiler or turbine. Another factor limiting chromia evaporation in modern advance steam cycles is that ferritic steels are extensively used, and they do not form chromia scales. So the activity of Cr is lower in the scales, which from Eq. 1 lowers the partial pressure of $CrO_2(OH)_2(g)$, which in turn decreases chromia evaporation (Eq. 9).

6. MITIGATION

As discussed in the Hypothetical Superheater-Steam Pipe-HP Turbine Steam Path section, relatively small changes in chromia activity at the oxide-steam interface can greatly influence the amount of chromia evaporation. For the alloys of interest, which contain Ti and/or Mn, large evaporation rates may be self-correcting as chromia is lost from the surface and concentrates Ti and Mn oxides (and thus reduces the chromia activity). Decreases in chromia activity along the steam path can result in little or no evaporation if the steam becomes saturated in $Cr_2(OH)_2$ relative to the lower surface $PCr_2(OH)_2$that arises from the lower aCr_2O_3.

Another mitigation would be the use of a non-chromia forming coating. An example would be a thermal barrier coating (TBC), such as one based on yttria stabilized zirconia (YSZ). Such coatings are being postulated for use to reduce turbine blade erosion. This would eliminate the transport path for chromia evaporation. Even if the coating were porous, it would effectively increase the gaseous diffusion boundary layer to the thickness of the coating, thus eliminating much of the detrimental gas velocity effects.

7. CONCLUSIONS

A methodology was developed to calculate Cr evaporation rates from Cr_2O_3 with a flat planar geometry (Holcomb, 2008), was expanded upon to allow for interior cylindrical geometries, and to allow for the effects of $CrO_2(OH)_2$ saturation within the gas phase (Holcomb, 2009). This approach was combined with Cr diffusion calculations within the alloy (with a constant flux of Cr leaving the alloy from evaporation) to predict Cr concentration profiles as a function of exposure time and to predict the time until the alloy surface concentration of Cr reaches zero.

An important aspect of chromia evaporation in alloys of interest (that contain Mn and Ti) is the possibility of reduced chromia activity at the metal surface. A reduction in chromia activity will reduce chromia evaporation. It was found that increasing the gas flow rate led to increased chromia evaporation and decreased apparent chromia activity. Increasing the water content in the moist air increased the evaporation, but results were mixed with its effect on chromia activity.

A hypothetical super heater (SH) tube, steam pipe, and high pressure (HP) turbine steam path was assembled and examined with the methodology. At the U. S. Department of Energy's goals of 760° C and 340 atm, the time until breakaway oxidation was predicted to be quite short for the turbine blade, and of concern within the steam pipe and the higher temperature portions of the SH tube. The predicted time until breakaway oxidation increases dramatically with decreases in temperature and total pressure. However, it was shown that chromia activity variability within the steam path (i.e., chromia activity in the HP lower than that in the SH/steam pipe) could reduce or eliminate chromia evaporation. The steam could become saturated in $Cr_2(OH)_2$ relative to the lower partial pressure of $Cr_2(OH)_2$ at the metal surface (lowered by the lower chromia activity). Thus the higher expected chromia evaporation rates in the HP section could be self-mitigating as the loss of chromia from the scale, which increases the relative amounts of Ti and Mn oxides) reduces the chromia activitiy at the oxide surface.

REFERENCES

1. H. Asteman, J.-E. Svensson, L.- G. Johansson, M. Norell, 1999 Indication of Chromium Oxide Hydroxide Evaporation during Oxidation of 304L at 873 K in the Presence of Water Vapor. Oxidation of Metals, 52 1-2, 95 111 , 0003-0770X.

2. H. Asteman, J.-E. Svensson, M. Norell, L.-G. Johansson, 2000 Influence of Water Vapor and Flow Rate on the High Temperature Oxidation of 304L; Effect of Chromium Oxide Hydroxide Evaporation. Oxidation of Metals, 54 1-2 11 26 , 00030-770X.

3. C. W. Bale, A. D. Pelton, W. T. Thompson, G. Eriksson, K. Hack, P. Chartrand, S. Decterov, J. Melançon, S. Petersen, 2007 FactSage 5.5, Thermfact and GTT-Technologies (2007).

4. R. B. Bird, W. E. Stewart, E. D. Lightfoot, 1960 Transport Phenomena, John Wiley & Sons, 0-471-41077-2 New York.

5. E. Essuman, G. H. Meier, J. Żurek, M. Hänsel, W. J. Quadakkers, 2008 The Effect of Water Vapor on Selective Oxidation of Fe-Cr Alloys. Oxidation of Metals, 69 3-4 143 162 , 0030-770X.

6. E. N. Fuller, P. D. Schettler, J. C. Giddings, 1966 Industrial & Engineering Chemistry, 58 5 19 27 , 1226-086X

7. D. R. Gaskell, 1992 An Introduction to Transport Phenomena in Materials Engineering, Macmillan Publishing, 0-02-340720-4New York.

8. C. Gindorf, K. Hilpert, L.L. Singheiser, 2001 Determination of Chromium Vaporization Rates of Different Interconnect Alloys by Transpiration Experiments, In: Solid Oxide Fuel Cells (SOFC VII), Yokokawa, H. & Singhal, S.C. (Eds.), PV 2001-16, 793 802 , The Electrochemical Society Proceedings Series, 978-1-56677-322-5 Pennington, NJ.

9. H. C. Graham, H. H. Davis, 1971 Oxidation/Vaporization Kinetics of Cr2O3 . Journal of the American Ceramic Society, 54 2 89 93 . 0002-7820

10. Haynes International. 2008 a Haynes 230 Alloy, H-3135C, Haynes International, Kokomo, IN.

11. Haynes International. 2008b Haynes 282 Alloy, H-3172A, Haynes International, Kokomo, IN.

12. G. R. Holcomb, 2008 Calculation of Reactive-Evaporation Rates of Chromia. Oxidation of Metals, 69 3-4 163 180 , 0030-770X.

13. G. R. Holcomb, 2009 Steam Oxidation and Chromia Evaporation in Ultrasupercritical Steam Boilers and Turbines. Journal of the Electrochemical Society, 156 9 C292 C297 . 0013-4651

14. G. R. Holcomb, D. E. Alman, 2006 Effect of Manganese Additions on the Reactive Evaporation of Chromium in Ni-Cr Alloys, Scripta Materialia, 54 10 1821 1825 , 1059-9495

15. G. R. Holcomb, D. E. Alman, Ő. N. Doğan, J. C. Rawers, K. K. Schrems, M. Ziomek-Moroz, 2007 Steam Turbine Materials and Corrosion. Proceedings of the 21st Annual Conference on Fossil Energy Materials, Knoxville, TN

16. F. P. Incropera, D. P. De Witt, 2001 Fundamentals of Heat and Mass Transfer (5th edition)., John Wiley & Sons, 978-0471386506 New York.

17. O. Kubaschewski, C. B. Alcock, 1979 Metallurgical Thermochemistry (5th edition), Pergamon Press, 0080221076New York.

18. F. Masuyama, 2001 History of Power Plants and Progress in Heat Resistant Steels, ISIJ International, 41 6 612 625 , 0915-1559

19. G. H. Meier, K. Jung, N. Mu, N. M. Yanar, F. S. Pettit, J. P. Abellán, T. Olszewski, W. J. Quadakkers, G. R. Holcomb, 2010 Effect of Alloy Composition and Exposure Conditions on the Selective Oxidation Behavior of Ferritic Fe-Cr and Fe-Cr-X Alloys. Oxidation of Metals, 74 5-6 319 340 , 0030-770X.

20. E. J. Opila, D. L. Myers, N. S. Jacobson, I. M. B. Nielsen, D. F. Johnson, J. K. Olminsky, M. D. Allendorf, 2007 Theoretical and Experimental Investigation of the Thermochemistry of CrO2(OH)(2)(g). Journal of Physical Chemistry A, 111 10 1971 1980 , 1089-5639

21. A. R. Paul, K. N. G. Kaimal, M. C. Naik, S. R. Dharwadkar, 1994 Lattice and Grain Boundary Diffusion of Chromium in Superalloy Incoloy-800. Journal of Nuclear Materials, 217 1-2 75 81 , 0022-3115

22. X. Peng, J. Yan, Y. Zhou, F. Wang, 2005 Effect of Grain Refinement on the Resistance of 304 Stainless Steel to Breakaway Oxidation in Wet Air. Acta Materialia, 53 19 5079 5088 , 1359-6454

23. F. S. Pettit, G. H. Meier, 2006 Fundamental Studies of the Durability of Materials for Interconnects in Solid Oxide Fuel Cells, Final Report, DOE Award: DE FC26 02NT41578.

24. B. S. Petukhov, 1970 Heat Transfer and Friction in Turbulent Pipe Flow with Variable Physical Properties, In: Advances in Heat Transfer, 6 Irvine, T.F. & Hartnett, J.P. (Eds.), 503 564 , Academic Press, 978-0120200061 New York.

25. Royal Institute of Technology, Foundation of Computational Thermodynamics. 1999 MOB2, Mobility Database, Version 2.0, Royal Institute of Technology, Foundation of Computational Thermodynamics, Stockholm, Sweden.

26. N. Saunders, 2000 Ni-DATA, Version 7, Thermotech Ltd, Surrey, UK.

27. J. C. Slattery, R. B. Bird, 1958 American Institute of Chemical Engineers Journal, 4 2 137 142 , 0001-1541

28. Special Metals Corporation. 2004a Inconel Alloy 740, SMC-090, Special Metals Corporation, Huntington, WV.

29. Special Metals Corporation. 2004b Nimonic Alloy 263, SMC-054, Special Metals Corporation, Huntington, WV.

30. Special Metals Corporation. 2005 Inconel Alloy 617, SMC-029, Special Metals Corporation, Huntington, WV.

31. Special Metals Corporation. 2006 Inconel Alloy 625, SMC-063, Special Metals Corporation, Huntington, WV.

32. S. C. Stultz, J. B. . Kitto, Eds 1992 Steam (40th edition), Babcock & Wilcox, 0963457004Barberton, Ohio.

33. C. S. Jr Tedman, 1966 The Effect of Oxide Volatilization on the Oxidation Kinetics of Cr and Fe-Cr Alloys. Journal of the Electrochemical Society, 113 8 766 768 , 0013-4651

34. Thermo-Calc Software AB. 2006 Dictra, Diffusion Simulation Software, Version 24, Thermo-Calc Software AB, Stockholm, Sweden.

35. R. Viswanathan, J. F. Henry, J. Tanzosh, G. Stanko, J. Shingledecker, B. Vitalis, R. Purgert, 2005 U.S. Program on Materials Technology for Ultra-Supercritical Coal power Plants. Journal of Materials Engineering and Performance, 14 3 281 292 , 1059-9495

36. D. P. Whittle, D. J. Evans, D. B. Scully, G. C. Wood, 1967a Compositional Changes in the Underlying Alloy during the Protective Oxidation of Alloys. Acta Metallurgica. 15 9 1421 1430 , 1359-6454

37. D. P. Whittle, G. C. Wood, D. J. Evans, D. B. Scully, 1967b Concentration Profiles in the Underlying Alloy during the Oxidation of Iron-Chromium Alloys. Acta Metallurgica, 15 11 1747 1755 , 1359-6454

38. R. H. S. Winterton, 1998 Where did the Dittus and Boelter Equation Come From? International Journal of Heat and Mass Transfer, 41 4-5 809 810 , 0017-9310

39. A. I. G. Wright, P. F. Tortorelli, 2007 Program on Technology Innovation: Oxide Growth and Exfoliation on Alloys Exposed to Steam, Report 1013666, EPRI, Palo Alto, CA.

40. D. J. Young, B. A. Pint, 2006 Chromium Volatilization Rates from Cr2O3 Scales into Flowing Gases Containing Water Vapor. Oxidation of Metals, 66 3-4 137 153 , 0030-770X.

8

CHAPTER

Thermodynamics of a Shallow Solar Still

Juan Cristóbal Torchia-Núñez[1], Jaime Cervantes-de-Gortari[2], Miguel Angel Porta-Gándara[3]

[1]*Independent Energy Consultant, San José, Costa Rica*

[2]*Mechanical Engineering Department, National University of Mexico, Mexico City, Mexico*

[3]*CIBNOR, La Paz, BCS, Mexico*

ABSTRACT

A detailed summary of the most relevant aspects of the thermodynamics of a shallow solar still is presented, including historical features not often found in the literature. Solar distillation has grown from applying empirical knowledge to advanced modeling and simulation. Geometrical, environmental and operational parameters of the solar still to heat transfer phenomena including evaporation and condensation, are taken into account in this overview, giving a comprehensive structure and classification to the study of solar stills from the thermodynamic point of view. The article describes global parameters, such as solar radiation, wind speed and thermal insulation among others and how they have been taken into account in the literature. Also, a distinction between internal and external heat transfer phenomena is proposed for clarification. Exergy balance is included to account for thermodynamic imperfections in the several processes inside the solar still.

Keywords: Solar Still, Overview, Thermodynamics, Modeling, Exergy, Heat Transfer

1. INTRODUCTION

1.1. History of Solar Distillation

Solar distillation of sea water is without doubt, one of the few possibilities to satisfy the water needs in many regions of the world in a sustainable way. It is an ancient process that human beings have learnt in order to satisfy different needs. Aristotle clearly explained in his writings the evaporation process as part of the water cycle in nature and the ancient Egyptians showed solar distillation in their paintings and hieroglyphics.

Full accounts of solar stills within the context of seawater desalination have been presented [1] where an extensive description of various designs of solar stills, from the single-box still to the sophisticated mutiple-effect distiller and greenhouse-inclusive types are described. These authors mentioned that direct solar desalination systems had low operating and maintenance costs, but required large installation areas and high initial investment. However, this is an appropriate solution for remote areas and small communities in arid and semi-arid regions lacking water. These authors[1] pointed out that the first conventional solar still plant was built in 1872 by Charles Wilson in the mining community of Las Salinas in Northern Chile. This still was a large basin-type still used to supply fresh water from brackish feed water to the community, with a total capacity of about 23 m^3/day and lasted 40 years until the mines were exhausted. A notable historical account of this project can be found in [2] , which offers a social and historical background about Wilson, the solar distillation plant and other authors who wrote about it.

Few other historic reviews with different emphases have been reported [3] for designs and experiences during the 50 s to 70 s; and more recently, an extensive account of solar stills has been published [4] .

Solar distillation is one of the oldest methods used to produce fresh water for different basic human needs. According to [5] and [6] , one of the first treatises about solar distillation is De distillationelibri IX, written by G.B. della Porta in 1589. This work describes a solar distillation with glass hemispheres directly exposed to the sun. In [7] mentioned Lavoisier and its method of solar concentrators for heating alembics in 1770. It is also described in [7] , manufacturing and operation advances of conventional basin and portable solar stills during WWII. This author analyzes some isolated elements of the solar still independently, showing different experimental results, such as the efficiency and fresh water production as a function of solar energy.

Perhaps the most important work of the last decades is [8] . This is the first work in which the basis of a complete analysis of the energy balance applied to the different elements of the solar still is presented. The analysis results in a lumped-parameters set of equations for simple and multiple-effect solar stills. The experimental work is based on five "diffusive" solar stills in operation for several months, where an electrical heater supplies energy to the solar still instead of solar radiation. The correspondence between experimental and calculated distillate production results is not good [8] ; however, the model is found as a proper mathematical tool for design purposes. Furthermore, [8] presents a general energy efficiency for the still defined as the ratio of heat used for evaporation in comparison with the amount of energy supplied to the still, reporting an efficiency of 57.7%.

Morse and Read [9] presented an important theoretical and experimental study in the solar distillation literature due to the parametric nature of thermal analysis and the method used to solve it. The method consists of a characteristic chart where the different heat fluxes can be expressed as a function of the glass cover temperature. Parameters involved in this study are wind velocity, ambient temperature and heat losses from the base.

Cooper [10] presents a purely theoretical, finite-difference model for the energy balance of the glass cover and the brine. The influence of several parameters, such as initial mass of brine—with and without thermal insulation under the basin—, is analyzed. Several other later works ([11] - [15]) have established the lumped-parameter equations for the energy balance, based on the model in[8] with several useful modifications.

In the authors' opinion, Solar Distillation [6] is one of the most important and complete works in the Solar Distillation field. It describes the heat transfer model of a solar still, the energy balances and the results of distillation yield as a function of time. Also, there is an extensive description of systems, configurations and applications of solar stills along with certain aspects of engineering economics.

1.2. Description of a Single-Basin Solar Still

The simplest design of a solar still consists of a rectangular box with a transparent upper cover as inFigure 1. This cover is usually a glass or a plastic sheet where, in the inner face, condensation is produced. The side walls are often made of an insulating and impermeable material to reduce heat and vapor losses. Solar stills can be mounted above the floor or lying plainly on the ground; either case, the bottom of the still is specially covered with an insulating material to reduce heat losses that can be particularly high along the direction normal to the bottom.

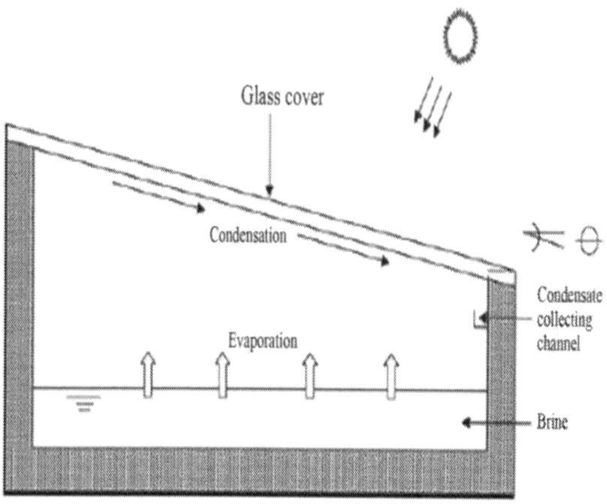

Figure 1. Schematic view of a single-basin solar still.

The solar distillation process is as follows: 1) the still is partially filled with brine in the bottom deposit which is a black surface (collector) used to absorb incoming radiation after it passes through the glass cover and the brine; 2) the collector increases its temperature and transfers heat to the brine; 3) water evaporates at the free surface; 4) a natural convection flow of humid air circulating inside the enclosure takes place due to the temperature difference between the free surface of the heated brine and the upper cool cover; 5) this inclined transparent sheet serves as a condensing plate where the distillate water runs by gravity along its internal face to a small collector channel in the shortest sidewall of the arrangement. There are other several designs for solar stills varying in complexity and applications. A thoughtful and extensive review of solar stills designs and their applications can be found in [1] [6] [16] .

2. GLOBAL PARAMETERS OF A SOLAR STILL

We call Global Parameters to characteristic properties, dimensions and factors that influence the distillation yield sprouting from climate, configuration and designing decisions. These parameters are usually studied by researchers in the solar still field to understand the influence on overall still variables such as distillation yield or efficiency. There are other several parameters that could be taken into account into a list of this type; however, the authors feel that these are representative for introduction purposes in this study.

2.1. Solar Radiation

The Sun is in essence a huge fusion nuclear reactor where the gravity pushes together hydrogen atoms at an enormous pressure and transforms them into heavier helium atoms. This process produces a deficit in the mass balance between hydrogen and helium, and this lost mass is actually the energy produced by the fusion reaction and radiated away. This is what we call solar radiation.

Solar radiation represents the driving force that keeps a solar still running and producing distillate water. Some 1360 W/m^2 of radiation power reaches the Earth's atmosphere. This is a rough value called "The Solar Constant", which actually is not constant because it varies some 3% due to variations in Sun-Earth distance along the year. Also, there are periodic variations in the power emitted by the Sun as the measurements from Spatial Agencies' satellites and Observatories around the planet have shown.

Some 30% of this radiation is reflected by the atmosphere so that, let us say, 800 W/m^2 reaches the surface of the Earth in a very sunny day near the Equator. This means that, if completely absorbed, 1 m^2-surface collector receives the same power than eight 100-Watts light bulbs. In just 12 hours, this 1-m^2 device receives almost 10 kWh of solar energy, that is, 10 days of a color TV's usage. According to the World Bank data [17] in average, a person in the US consumed 36.29 kWh of electricity each day back in 2011. A perfect solar collector would need just 3.5 m^2 to satisfy the electricity of one person by using 12 hours of sunlight.

Now, these are optimistic numbers because there are geographical, seasonal and weather variations that reduce the 800 W/m^2 down to some 170 W/m^2 as an average according to the World Energy Council [18] . This figure is averaged in time (24 hrs) and across regions of the planet. There are places on Earth that far exceed 170 W/m^2, especially those near the Equator. Furthermore, no solar device has 100% efficiency; thus, the power output of a device using solar radiation would be lower than 170 W/m^2. Solar radiation is a resource that needs more efficient conversion devices and relative large areas in order to compete with fossil and fissile energy sources. Although this work is not dedicated to the resource of solar radiation and its characteristics, it has to be noted that the discussion of solar energy conversion has to be centered on the efficiency of devices, instead of the solar resource. The solar resource is very little negotiable, unless solar concentration steps into the discussion. There is not much to do about the amount of energy reaching the surface of the Earth but there is a lot to do with the amount and the quality of solar devices can be placed in the surface of the Earth. Most of the countries where solar conversion devices are in full development are located in poor solar-radiated zones like Northern Europe which shows that the scarce solar radiation should not be a decisive factor when considering options about the energy supply.

2.1.1. Description of the Radiation Phenomenon

Radiation is a mechanism of heat transfer between two bodies. The main feature of radiation lies in the fact that there is no need of a participating medium between the source of the radiation and a receiver. Radiation can transport heat through vacuum and the main example is solar radiation going through an almost empty space between the Sun and the Earth. As long as radiation travels through space, whether it is empty or not, it can be described accurately as a wave. This wave is produced by vibrations in the particles of the source. Each time a molecule changes its energy levels, radiation is produced [19] . However, radiation also behaves as particles called photons.

In summary, radiation can be both, a wave and a particle. For some particular conditions it will behave like a wave (the interference phenomenon) and for some other conditions it will be described as a particle (the photoelectric phenomenon). This duality has intrigued and mesmerized physicists since it was first proposed.

The nature of radiation is complex and it basically needed a whole new theory just to be completely understood and explained. One part of the knowledge about radiation was given by Electromagnetic Theory, which was completed when James Clerk Maxwell established the link between the laws of electricity and magnetism. However, there was a situation in the theory that was dramatically called Ultraviolet Catastrophe where experiments showed that the mathematical model used to describe the ultraviolet power radiation was wrong for certain frequencies. Planck solved the problem by bringing up something that was later called Quantum Theory [20] .

Radiation is a phenomenon that depends on several parameters:

1) Intensity;

2) Wavelength;

3) Angle of the ray with respect to the normal of the surface of the source;

4) Optical properties of the material;

5) Solid angle described by the bundle of rays;

6) Temperature of the source and the receiver.

So it is easy to note that the modeling of radiation from the viewpoint of heat transfer is complex and multidependent. A unique characteristic of radiation is that the heat transported is proportional to T^4, different from the convection-conduction linear dependence on temperature. That is the reason why radiation becomes the dominant heat transfer mechanism when temperature increases as it can be perceived in incandescent bulbs, furnaces, combustion devices and, of course, the Sun, among others.

When it comes to solar conversion technology, the solar radiation values takes into account the solid angle produced by the solar disk in space, the integration of all solar wavelengths, and supposed to reach the solar still at a 90° angle from the horizontal. About the optical properties of a solar still, the glass cover is considered a surface with very high transmittance and no absorptance. The brine, if considered as part of the analysis, shares almost the same optical characteristics than the glass cover although care should be taken whenever the level of brine or its brackishness increases. The bottom of the still is considered to be a perfect absorber with zero transmittance and reflectance. The air-vapor mixture within the solar still is simply disregard as a non-participating medium. There is a vast literature about radiation [19] and solar radiation for engineering purposes [21] where the reader can find complete information regarding these topics.

2.1.2. Direct, Diffuse and Global Solar Radiation

Radiation can be classified into three types that are especially important in solar conversion technology: direct, diffuse and global.

Direct solar radiation is the part of the radiation that comes straight from the sun without being scattered by any participating medium. It can be described as a coherent pencil of rays that does not diverge from its original path. It can be measured by a pyrheliometer. A pyrheliometer is a device that uses a collimator tube which is blackened and diaphragmed to receive the ray and drives it into the sensor. The collimator tube allows only an angular acceptance of solar radiation of about 5° and 6° of the total sky hemisphere [22] . The sensor is a set of thermopiles located at the bottom of the collimator tube. As the thermopiles receive solar radiation, they increase in temperature and produce an electrical current which is a linear function of solar radiation intensity.

Diffuse radiation is constituted by electromagnetic waves that have been scattered by matter within the path of the wave so many times that its intensity, wavelength and directional properties are completely different from those of the radiation leaving the source. It is the radiation coming from the sky hemisphere that is not direct radiation. Pyranometers are instruments that measure diffuse radiation or global radiation depending on how the instrument is set. A pyranometer is a device that uses a transparent hemisphere glass at the top that allows radiation to heat, again, a thermopile that produces an electrical current when radiation lies within 0.3 and 2.8 μm which is 99% of solar radiation [22] . However, some pyranometers might have silicon photovoltaic sensors. For diffuse radiation measurements, no direct radiation is allowed into the pyranometer, therefore, there are disks or shadow bands located above the glass hemisphere to block and then prevent direct radiation reaching the pyranometer's sensor. Corrections are needed for accurate measurements of diffuse radiation when using a pyranometer due to the blockage.

Global radiation is obtained simply by adding the direct and diffuse radiation components. Global radiation is measured with a pyranometer too, the difference being that there must be no objects between the hemisphere and the instrument. Both pyrheliometers and disks or shadow bands for diffusion radiation pyranometers need a tracking system to follow the sun's path across the sky both for daily and seasonal variations.

2.1.3. Maps and Units

As any other solar energy conversion technology, solar stills rely on the intensity of solar radiation as the most critical and important parameter. With the advent of solar energy engineering devices and concerns over climatic changes around the world, solar radiation maps have become available in almost every region on the planet. These maps are color or gray tones representations that show the amount of solar resource reaching the ground as a function of space (regions) and time (per day or per year). It can be seen in this maps that the higher intensity radiation is distributed around the Equator and especially on desert areas of the planet such as Australia, Mexico, Middle East and North Africa.

It is important to clarify some of the terms and units employed in solar energy conversion technology since there are many in the literature to describe different concepts interrelated. We suggest following the nomenclature given in [23] . Solar radiation is simply a general way to describe the radiation phenomenon that stems from the sun. Irradiance is the radiation power per unit area and its units are W/m^2. Irradiation is the radiation energy per unit area and its units are J/m^2. The difference between these two magnitudes is time.

Irradiance is measured instantaneously. Take, for example, that at 6 o'clock in the morning, an instrument measures an irradiance value of, say, 100 W/m^2 but two hours and a half later, the instrument reads 345 W/m^2. One instant is for only one measure that results in only one value. On the other hand, an instrument that measures irradiation needs an interval of time. From, say, 6:00 a.m. to 8:00 a.m., the accumulated value of energy that arrived as radiation is, let's say, 1500 J/m^2. It is a measurement that needs some time to be taken; it is not possible to have an instantaneous value of irradiation. The difference between the two is simply a matter of the information they provide: an instantaneous measure of the energy in radiation (Power in kW/m^2) or the accumulated energy in a certain time interval (Energy in kWh/m^2 or J/m^2).

When it comes to Solar Maps, units are not the same for all of them since they are arbitrarily chosen by the research group or editorial board of the publication. However, they can be easily transformed between one another and both are clear information about the solar resource. Most of the maps show Irradiance as the variable of study, because they use kWh/m^2 per year or per day, that is the time-averaged energy (kWh) falling in some region.

2.2. Thermal Insulation

The goal of thermal insulation is to minimize heat leaks through the side and bottom walls of the solar still. Thermal insulator materials have to fulfill the requirements of a long-life operation while resisting aggressive environments such as humid and saline near the sea coasts or arid and corrosive in desert industrial zones keeping in mind that they should be as cost-effective as possible.

Thermal insulation is a resistance to heat fluxes that depends on the thermal conductivity of the material that surrounds the brine. Definition of thermal conductivity comes along with Fourier's constitutive relation that models heat transfer phenomenon by conduction through a medium.

If the thermal conductivity is reduced, the temperature difference between the collector and its surroundings will increase, for instance, the still will be better insulated. Insulation is considered an isotropic and independent of temperature in order to be characterized by a constant thermal conductivity. Heat conduction losses through the insulation are reduced if ambient temperature increases or wind speed decreases (for a mounted still).The influence of insulation has been proved to be significant up to some degree. Theoretically [9] and experimentally [24] validate this tendency. It is seen that the usefulness of insulation manifests up to a certain value of thickness. Cooper [10] showed that the absence of a thermal insulator has no significant effect, but a 30% increase of distillate production is obtained for 0.5 to 12 inches of brine-glass cover gap using thermal insulator.

2.3. Wind Speed

Wind speed is the most unpredictable of all solar still parameters due to its very own nature. Solar stills are not necessarily located in regions where charts of wind speed and direction are known or available as is the case for regions where, for example, wind turbines are installed. Therefore, it is a parameter that cannot be controlled or specified with relative good accuracy for solar still operation.

The effect of wind speed is a reduction in the overall temperature of a solar still which would be an undesirable effect. However, if the wind flows over the glass cover it extracts heat from it and promotes an increase in still productivity due to a larger temperature gap between brine and glass cover. Looking into the detail of heat transfer phenomena within the solar still, the more heat we extract from the glass cover, the higher yield we would get out of it.

When the outside surface of the glass cover is blown with colder air, heat is dissipated to the surroundings. This dissipation is increased whenever wind speed increases due to a higher heat transfer coefficient. McAdams [25] presented an expression for the heat transfer coefficient over a surface as a function of speed. This expression have found widespread acceptance in the solar still literature but it has to be noted that the expression is an experimental result [25] found by Jurgesback in 1924 for a parallel airflow at room temperature over a vertical copper plate. These specific experimental conditions are far from representing a solar-still glass cover, however, we think that there wasn't much information regarding heat transfer coefficients back in the 1960's and McAdams [25] was a strong reference for heat transfer studies.

Wind speed is a parameter that has been studied with no conclusive results regarding a positive or negative influence on distillation. El-Sebaii [26] has numerically shown that at higher wind speeds, both temperature difference and distillation output increase but only up to a certain value. Its influence on distillation rate has been debated with contradictory results in several articles. Morse and Read [9] in their parametric study conclude that wind speed is not relevant for distillation output. Malik et al. [6] reported a steady increase in daily output as the loss coefficient decreased. Cooper [10] found that wind speed over 5 miles per hour have a marginal augmentation effect on the distillate yield.

2.4. Brine Inlet Temperature

Considering the heat transport phenomena, brine temperature is relevant to all heat fluxes from basin water, that is, evaporation, free convection and radiation. Evaporation rates increase for higher water temperatures due to the fact that less energy is needed to break the barrier from liquid to gaseous form. Another way to look at it: at some specific pressure (atmospheric pressure), water will evaporate at the saturation temperature and the higher brine temperatures are the closer brine is to saturation temperature. Therefore, it is not surprising that some techniques use residual heat to preheat brine ([27] -[29]).

In fact, solar stills are at the bottom of high temperature requirement for thermal devices needing only some 5°C to 10°C above environmental temperature in order to function properly, in contrast with metallurgical furnaces or power plants, for example. Most of the industries that produce residual heat don't have solar stills to be fed because they have full access to clean water by other means. Nonetheless, it would be interesting to promote the idea, whenever it is cost-effective, that solar stills could be a part of every industry if space or costs are not restricted. Distilled water could be for staff use or distributed to the community as a social benefit for the company or district.

For high latitude regions where cold brine is found and the possibility to preheat it is economically unfeasible, low-level-basin or wick-type solar stills should be considered for higher distillate yields.

2.5. Brine Depth

The amount of brine inside the basin determines the brine depth or the level of brine. This is a parameter that appears only in analyses where the heat transfer between the collector and the brine is considered. There are studies where the brine and the bottom of the still are thought as the same system [8] . This has two consequences: first, this means that radiation doesn't contribute to heat the brine; and second, that all the performance of the still is independent on the water depth which is arguable, as experimental results in the literature.

Water depth implies the total amount of water in a solar still. The bottom of the solar still has a surface which is fixed. The brackish water or brine that is poured into the basin spreads through all that area and reaches a certain level inside the basin of the solar still. The mass of water is directly proportional to the thermal inertia of a solar still. Thermal inertia is how resistant the brine is to change its temperature. It depends on the heat capacity and the mass of the brine. The higher the thermal inertia, the greater the time it is required for the still to produce distillate.

Tripathi and Tiwari [30] showed that for 0.10 m and 0.15 m of brine depth, higher distillation rates were achieved due to thermal storage effect. They found not only distillation was influenced by brine depth due to thermal storage effect but also due to the modification of the heat transfer coefficient between water surface and inner condensing cover by changing their average spacing.

Tayel et al. [31] studied different configurations and parameters in solar still productivity, one of them being brine depth. The study shows that the highest distillation rate is found for a 2-cm brine depth. Badran [32] experimentally studies enhancement parameters for a solar still. As brine depth increases, productivity decreases due to an increase in heat capacity of the water in the basin which, in turn, reduces evaporation rate for the same solar input. Furthermore, in terms of how solar radiation is affected by a layer of water depending in its depth is reported by Kreider and Kreith [33] .

2.6. Cover Tilt Angle

The cover tilt angle is fundamental for the production of distillate water. It provides the proper inclination for the condensate to flow out of the still: a small angle would produce condensate to drip from the interior face of the glass cover while a large angle would increase still costs and compromise mechanical support.

As Rahbar and Esfahani [34] pointed, the model from [8] has a limitation due to the 10°-angle value for the cover tilt, which is a small value even for low-latitude locations. This limitation has promoted a considerable amount of studies regarding cover tilt angle in solar stills, whether they are single or double sloped.

A thorough survey of studies where cover tilt angles were examined can be found in Khalifa [35] . This author examined the relationship between cover tilt angle and latitude for optimal distillate production. It seems that for higher productions, cover tilt angle and latitude should be close to each other, which is a rule of thumb for solar water heaters. However, conflicting results can be found about this parameter as the author pointed out.

3. THERMAL BALANCES

Current trends on solar distillation are based on new geometrical designs, evaporative-enhancing materials, its coupling with photovoltaic cells and surfactants supplied into the brine. But in the opinion of the authors, there has been little effort in trying to explain and widen fundamental knowledge on transport phenomena, primarily inside the still, where the useful effect occurs. Evaporation and condensation are critical phenomena for solar distillation and most of the theoretical and experimental relations are simplifications leading to very useful design information. However, in this work emphasis is made in reviewing classical aspects of heat and mass transfer and exploring solutions based on ideas located in other fields of study.

A series of thermal analyses considering the transient operation of the still, are presented, based on the published literature and the author's own experience. These include physical and mathematical modeling of internal and external transport phenomena that affect the response of the still, as well as thermodynamic balances of the whole equipment and environment coupled systems. The internal analysis comprehends natural convection from basin to brine and from brine to air, and thermal radiation from the basin to the glass cover. The external analysis considers solar radiation from the glass cover to the surroundings and into the basin.

3.1. Internal Transport Phenomena

3.1.1. Natural Convection from Brine to Glass Cover

Convection is the method in which heat flows by the motion of a fluid. Natural convection is the heat flow when the fluid is set in motion by some force inside the fluid. This force is produced by a difference in density within the fluid. Density changes may be produced by a temperature difference or a concentration difference. Natural convection is a critical mechanism inside the still: no water steam evaporating from the brine could reach the condensation surface (glass cover) without the motion of natural convection cells.

For many years it was taught that energy and mass transport in solar stills were mainly accomplished by diffusive transfer in most part of the still, with a little boundary layer type flow close to the glass cover and may be in a small region adjacent to the liquid surface [8] . The reason for this interpretation can be found, perhaps, in that there were no detailed experimental and visualization studies except for the investigation conducted by Baum and Baimarov [36] , showing a physical structure resembling very much to the one described above.

Nevertheless, natural convection in a solar still has been recognized by contemporary authors, as the most important mechanism for heat transport along with evaporation from the surface of the brine to the glass cover ([6] [37] [38]). The extensively used model for free convection in a solar still, widely referred to in the literature, is the one first proposed in [8] and explained in detail in[6] .

Dunkle [8] proposed that the Nusselt number for free convection could be expressed as:

$$Nu = C\left(Gr * Pr\right)^{n} \tag{1}$$

where the constants C and n are obtained from experiments, with different values that depend on the flow regime either, laminar or turbulentLaminar flow $10^4 < Gr < 3.2\times10^5 \quad C = 0.21,\ n = 1/4$

Turbulent flow $3.2\times10^5 < Gr < 10^7 \quad C = 0.075,\ n = 1/3$

Malik et al. [6] considered a turbulent regime which cancels out the presence of a length scale in the Grash of number and this result is interpreted as if the heat transfer coefficient h_{cw} would not depend on the height of the still (the distance between the surface of the brine and the glass cover), contrary to what Porta-Gándara et al. [39] have reported. Hongfei et al. [40] on the other hand, considered a laminar regime in the solar still, based in their own experimental results. To add even more uncertainty in the regime that one would expect, PortaGándara et al. [39] , Rubio et al.[41] and Porta-Gándara et al. [42] presented flow visualization results for a transition regime in the same class of solar still, Figure 2. In addition to the turbulent regime for natural convection within the still, Malik et al. [6] proposed a limited range of temperatures for which the values of the physical properties of the fluid remain constant with the temperature. With the appropriate consideration of mass transport during evaporation they arrive to the generally accepted expressions:

$$q_{cw} = 0.884\left[T_w - T_g + \frac{\left(P_w - P_g\right)}{268.9\times10^3 - P_w}\left(T_w + 273\right)\right]^{1/3}\left(T_w - T_g\right) \tag{2}$$

and

$$h_{cw} = 0.884 \Delta T'^{1/3} \tag{3}$$

where T_g is the temperature of the glass cover.

Malik et al. [6] used a linear expression for the pressure based on the temperature. There is an exponential approximation of the pressure function in terms of temperature which tends to be more accurate [37] .

When mass transfer occurs between two substances with different molecular masses, such as the case of water into air, a modified Grashof number is required, as described in Malik et al. [6] . The modification of the temperature difference affects the Grashof number in the following way:

$$Gr = \frac{x^3 g \beta}{\nu^2} \Delta T' \tag{4}$$

Where

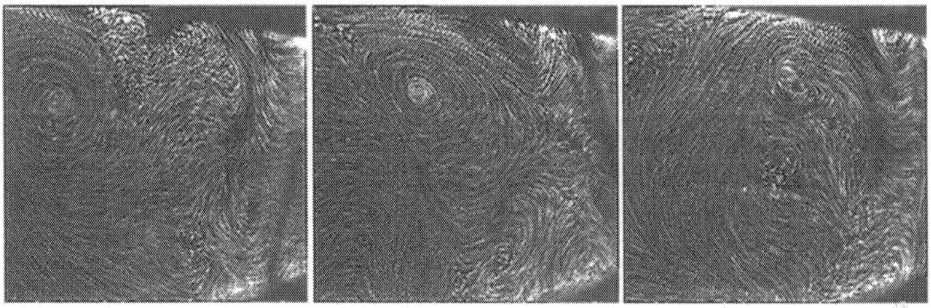

Figure 2. Visualization of fluid flow inside a solar still by Porta et al. [39] .

$$\Delta T' = \left(T_w - T_g\right) + \frac{\left(P_w - P_g\right) T_w}{2.8 - P_w} \tag{5}$$

Therefore, the Rayleigh number is

$$Ra = Gr \times Pr \tag{6}$$

where the Prandtl number is defined as

$$Pr = \mu c_p / k \tag{7}$$

In almost every heat transfer study, the Nusselt number is used as ratio of convection and conduction, describing how heat flux is transported between the walls into the fluid region.

$$Nu = h_{wg} x / k \tag{8}$$

3.1.2. Natural Convection from Basin to Brine

Once solar radiation increases the temperature of the basin, typically a black surface, heat is transported to the brine by natural convection. As a result, the brine is heated by convection currents from the inside bottom of the still and brine increases its temperature which, in turn, favors evaporation. In terms of modeling, this particular heat flux is often not taken into account by some authors, noticeably [8] . He establishes that the energy gain by the water is due to the absorption of the radiation of the still's base. This fraction is due to the high transmittance of, both glass cover and brine layer, and also due the high absorptance of the base. In this case, the solar still base and brine is the same thing for modeling purposes. Therefore, there is no basin-to-brine convection. Furthermore, the amount of water inside the basin may be so low that water convection cells may be inhibited and only conduction is present.

$$q_{colw} = h_{colw} \left(T_{col} - T_w \right) \tag{9}$$

3.1.3. Radiation from the Brine to the Glass Cover (or Elsewhere)

Radiation is perhaps the least dominant heat transfer phenomenon within the solar still due to the relative low temperatures of the overall solar still. Radiation from the brine to the glass cover can be divided in short and long wavelength. Short wavelength radiation is the amount of solar radiation that is reflected back from the brine surface or the bottom liner of the still to the glass cover. In most studies this radiation is simply neglected. On the other hand, long wavelength radiation (IR radiation) emitted from the basin to the glass cover is frequently taken into account.

The mathematical expression for heat radiation, in this case, is

$$q_{rw} = \sigma\varepsilon \left(T_w^4 - T_g^4 \right) \tag{10}$$

Theoretically, using a trial method, Dunkle [8] found that radiation accounted for 10% from the overall energy budget in the still when using a value of 0.9 for water emissivity.

3.1.4. Evaporation

Evaporation is the phase change of the brine or brackish water from liquid to gas and the mechanism that separates salt (or some other dissolved or suspended particles) from water, which is the first goal of any still: the separation process.

In the solar still, once the brine receives a certain amount of heat (enthalpy of evaporation) it changes from a liquid phase to a gaseous phase. Then it flows upward to the glass cover due to convective currents. The driving force for evaporation is a pressure difference. In the vast majority of cases, this pressure difference is produced by a temperature difference which modifies the partial pressure of steam in two regions of the still. In the brine surface, temperature is higher, so the partial pressure of steam, whereas inside the glass cover, partial pressure of steam is lower due to lower temperature.

Evaporation is a complex phenomenon that involves coupled heat and mass transfer, geometry parameters, external forces (gravity) and the properties of substances. From the viewpoint of the Kinetic Theory of Gases, evaporation is born from an unbalanced situation when more particles cross the interphase from the liquid to the vapour region than particles going from the liquid to the vapour region. Kinetic Theory calculates the mass flux from one phase to the other as a function of temperature and pressure [43]

$$j = \left(\frac{M}{2\pi R}\right)^{1/2} \frac{p}{\sqrt{T}} \tag{11}$$

Malik et al. [6] proposed a heat flux coefficient for evaporation in a solar still after the analysis of mass balance between the surface of the water and the glass cover. This coefficient is expressed as a function of the partial pressures of the vapor in the surface of the water and near the glass cover. The difference of partial pressures in these two regions of the still is the driving force that produces evaporation from the surface.

$$h_{ev} = \left(\frac{M_w}{M_a}\right)\left(\frac{\lambda}{C_{p,a}}\right)\left(\frac{1}{P_{tot}}\right) h_{cw} \tag{12}$$

After making some assumptions about the numerical values of water and air properties and pressure, the heat transfer coefficient for evaporation is:

$$h_{ev} = 0.013 h_{cw} \tag{13}$$

After some algebraic manipulation and considerations, it is found:

$$q_{ev} = 16.273 h_{cw} \left(P_w - P_g \right) \tag{14}$$

In general, heat released from evaporation is expressed by:

$$q_{ev} = \lambda \left(\frac{M_w}{M_a} \right) \left(\frac{h_{cw}}{C_{p,a}} \right) \left(P_w - P_g \right) \left(\frac{P_{tot}}{\left(P_{tot} - P_w \right)\left(P_{tot} - P_g \right)} \right) \tag{15}$$

3.1.5. Condensation

The condensation process in a solar still is a phase change from gaseous to liquid phase produced in the inside cold surface of the glass cover [44] . Unfortunately, most of the theoretical and experimental studies about solar stills almost completely disregard condensation as a fundamental part of the process of fresh water production. One notable exception is [45] where a model and experiments are developed in a tubular solar still in order to get correlations that take into account the effect of air in steam condensation.

At all lights, condensation is the final step towards distillate output in the sequence of processes within a solar still and it should be considered as a critical part in the distillation process, enhancement techniques and optimization-based design.

This topic has received a lot of attention in the condenser literature for power plants, specially nuclear power plants and refrigeration applications. However, it is important to note that industrial condensers work with just one component or species on its condensation surface (air is avoided, as a rule, due to corrosion problems) whereas the glass cover of the solar still condensates water in the presence of air, a non-condensable gas. In this regard, a great amount of literature has been produced in the nuclear engineering field to study the effect of noncondensable gases in condensation on vertical and horizontal tubes ([46] [47]).

In any case, condensation inside a solar still occurs in a boundary (interior face of the glass cover) where the fluid has small amounts of water vapor so that it shares this feature with psychrometry applications.

3.1.6. Diffusion

Diffusion is the transport of substance(s) within a region occupied by other substance(s) by means of molecular interactions. These substances are called species and in order to have diffusion, at least two species are needed in the same region. Diffusion is a very important subject of study for chemical engineering, kinetics, electrochemistry and other related analysis such as transport of biological components.

Mathematical modeling of mass transfer by diffusion can be approached in several ways ([43] [48]). However, most of the analyses use the constitutive relation called Fick's law to relate concentration gradients to diffusion mass fluxes. Water steam diffusing into air is a commonly known process and data such as diffusivity coefficients are easily available in the specialized literature. In a solar still, diffusion of evaporated water within air is neglected as an influential transport phenomenon inside a solar still in comparison to the mass transported by free convection currents.

Using scale analysis, the diffusion time needed for a particle of water steam to diffuse up to the glass cover is

$$t_{dif} = \frac{H^2}{D} \tag{16}$$

where H is the distance diffused by species and D, the diffusivity of the water vapor. Now, a 20-cm height from brine surface to the glass cover and 0.260 cm^2/s asthe value of diffusivity of water in air [48] , the diffusion time is approximately 1540 seconds or 26 minutes. On the other hand, the time for transport due to free convection is

$$t_{dif} = \sqrt{\frac{H}{\beta g \Delta T}} \tag{17}$$

Assuming the water vapor as an ideal gas, properties at 40°C and a temperature difference, ΔT equal to 20°C, the convective time is 1.64 seconds. In a solar still, diffusion is slow in comparison to convective transport. A 2-mm gap between brine surface and glass cover is needed for diffusion to be as dominant as free convection for mass transport. It is noticeable that [8] and [36] mentioned "diffusion stills" emphasizing diffusive process as the dominant driving force within a still.

Rahbar and Esfahani [34] included diffusion transport in their CFD study by adding the buoyancy force due to concentration gradients in the momentum equation. They presented contours of vapor mass fraction inside the solar still showing high concentration gradients near the surfaces and almost constant concentration regions in the bulk fluid along with the traces of convective cells, which is the result of a dominant convective transport.

3.2. External Transport Phenomena

As long as heat flows within the solar still, it is part of the Internal Transport phenomena. External Transport Phenomena are simply heat transfer losses from the solar still to the surroundings. From the efficiency point of view, some of these losses decrease the performance of a solar still so they need to be minimized. However, it is important to recognize that glass cover "needs" to lose heat so that a temperature difference between the brine and the glass cover is established in order to enhance distillate yield and perform properly. Thus, some heat losses should be minimized by design while other heat losses should not.

3.2.1. Conduction from Basin to Surroundings

Heat flows from every wall containing the basin to the surroundings by conduction, which is mathematically modeled by Fourier's law of conduction:

$$q_{ins} = -\frac{k\Delta T}{x} \tag{18}$$

Typically, cheap, insulating materials are used to build solar stills like wood, cement and fiberglass. For some fixed temperature difference between brine and ambient temperature, the largest surface conducts more heat than the others. This means that the basin bottom should be thicker than the other walls and less exposed to wind whenever possible. Conduction losses account for around 8% of energy input from solar radiation [8] .

3.2.2. Radiation Losses to the Surroundings

Basically there are two sources for radiation from the solar still: 1) Long wavelength radiation emitted from the basin and 2) short wavelength radiation reflected back by the glass cover.

There is a fraction of solar radiation that is reflected back to the surroundings due to the reflectivity of the glass cover and the walls of the still. A good solar still design would require the lowest possible value for reflectivity and absorptivity for the glass cover, so that most of the radiation is transmitted through the glass. Also, glass thickness should be kept as minimum as possible without compromising mechanical strength.

The radiation emitted by the glass cover to the surroundings

$$q_{ra} = \sigma\varepsilon\left(T_g^4 - T_a^4\right) \tag{19}$$

Due to the low emissivity of the glass and the low temperature gap between the glass cover and the surroundings, this term is a small fraction in comparison with the other heat fluxes.

3.2.3. Convection from Glass Cover to Surroundings

The dominant heat losses from the outer surface of the glass cover to the surroundings are by convection due to the wind. The expression for heat transfer by convection from glass to the environment is:

$$q_{ca} = h_{ca}\left(T_g - T_a\right) \tag{20}$$

It was already mentioned in Section 2.3 that there is an expression widely used in solar still literature for h_{ca} due to McAdams [25] using data of Jurges for a 0.5-m^2 copper plate:

$$h_{ca} = 5.7 + 3.8V \tag{21}$$

Duffie & Beckman [21] mentioned that probably this formula includes the effects of natural convection and radiation. Wattmuff et al. [49] on the other hand, found the expression

$$h_{ca} = 2.8 + 3V \tag{22}$$

which tends to give lower values for h_{ca}. Glass covers for solar collectors fits properly glass covers used for solar stills regarding materials and inclination angle, so these expressions may be used safely for calculations.

4. THERMODYNAMIC MODELING OF A SOLAR STILL

Thermodynamics is the science that studies states and their accessibility. States are a set of conditions (properties and external agents) that establish the description of a system completely. In Thermodynamics, many systems are described only by their pressure, temperature and specific volume. In a solar still, it is often seen that the knowledge of temperature and pressure suffices to characterize the state of any system within it, although more complex analysis should take chemical potential of steam and air into account.

Thermodynamics have proposed four laws that relate the possibility of a system reaching some particular states from an initial state or from a final state, tracking back its possible initial states. The Zeroth law is the recognition of a property called temperature, completely unidentifiable, elsewhere in science. The first law of Thermodynamics states the definition of energy and its consequence is the mathematical expression of the conservation of energy. The second law of thermodynamics was born as the result of the discovery that for a given amount of energy, it was impossible to use the whole of it for useful output.

Thermodynamics describes the evolution of systems in space and time from an initial to a final condition. Using its laws, Thermodynamics provides tools to find a maximum usage of energy to feed processes in order to satisfy several types of commodities, i.e. material, energetic, etc.

In a solar still there are several transport phenomena coupled in order to produce fresh water. Keeping track of each one of these fluxes and studying them under the scope of Thermodynamics, that is, defining systems and describing their processes using its laws, gives the opportunity to design better solar still systems, produce more fresh water or optimize the solar resource for a given solar still. Furthermore, the second law of thermodynamics establishes a limit in the amount of useful energy that can be extracted from every process. Knowing thermodynamics, among other things, equals to know the best theoretical thermal device for a given purpose and compare it with the real device.

4.1. Modeling of a Solar Still

In a solar still, part of the complexity of modeling lies in the fact that the three heat transport mechanisms appear within the solar still. Conduction and convection are proportional to a temperature difference. Meanwhile, radiation heat transfer is directly proportional to the fourth power of temperature. Another part of the complexity is that heat transfer convection mechanism depends highly on geometry, fluid/wall properties and temperature.

In the resulting mathematical model after a first-law analysis, the unknowns of the energy equations are temperatures. Temperatures are obtained once the input functions i.e. solar irradiation and ambient temperature and parameters are set.

The energy equations (see below) of the solar still are nonlinear, time dependent, ordinary differential equations. They are the result of a lumped parameter analysis, that is, variables do not change spatially. The nonlinear character is due to the heat transfer convective coefficients that depend on the temperature. The heat transfer coefficients are brought into the equation in order to model heat fluxes. Thermodynamics is unable to find relations to calculate heat fluxes. Transport phenomena are non-equilibrium phenomena; therefore, Classical Thermodynamics is unable to find a mathematical expression for heat, which is the study of Heat Transfer.

The expressions for solar irradiance and ambient temperature functions for a 24 hour period are, respectively:

$$G = \frac{G_{\max}}{2}\left[\sin\left(\frac{\pi t}{12}\right)\right] + \left|\left(\frac{G_{\max}}{2}\right)\left[\sin\left(\frac{\pi t}{12}\right)\right]\right| \tag{23}$$

$$T_a = A\sin\left(\frac{\pi t}{24}\right) + T_{a,i} \tag{24}$$

where t is the time in hours and $T_{a,i}$ is the ambient temperature at the beginning of the day in °C.

Equation (23) is the input of the solar still model, the driving force.

Assumed in the model is the following:

- All physical properties of materials are not affected by temperature differences;
- Vapor-air mixture and insulation are not regarded as systems or participating media;
- Brine and glass cover do not interact with incoming solar radiation;

- The shape factor value between the collector and glass cover is 1;
- The solar still is a closed system;
- There are no temperature gradients across brine depth, collector plate or glass cover.

Solar irradiance and ambient temperature are two time-dependent parameters with an important role in the transient energy balance of the still. Malik et al. [6] used Fourier series for modeling the periodic behavior of solar parameters.

4.2. Energy Balance

Energy is an inherent property of matter, so general that it is almost impossible to define. We do not quite perceive it, only its manifestations. There is nothing in the universe happening that cannot be counted as an energy transformation. Every imaginable process has to be with the conversion of one type of energy to another. In Thermodynamics, the First Law is a statement that defines a property called energy. A consequence out of the First Law of Thermodynamics is the Conservation of Energy.

Conservation of energy can be thought as an accounting exercise that basically establishes that what goes in equals what goes out plus the accumulation or minus the lost. The mathematical expression of the conservation of energy is one of the thermodynamic tools to study systems and several of its forms can be found in the vast literature of Thermodynamics.

Applying the conservation of energy to the solar still results on a three-equation system, for the glass cover, brine and collector, respectively, we obtain:

$$\underbrace{q_{ev} + q_{cw} + q_{rw} + a_g G}_{\text{in}} - \underbrace{\left(q_{ca} + q_{ra}\right)}_{\text{out}} = m_g c_g \frac{\mathrm{d}T_g}{\mathrm{d}t} \tag{25}$$

$$\underbrace{a_w G + q_w}_{\text{in}} - \underbrace{\left(q_{ev} + q_{cw} + q_{rw}\right)}_{\text{out}} = m_w c_w \frac{\mathrm{d}T_w}{\mathrm{d}t} \tag{26}$$

$$\underbrace{a_{col}G}_{\text{in}}-\underbrace{\left(q_{colw}+q_{ins}\right)}_{\text{out}}=m_{col}c_{col}\frac{\mathrm{d}T_{col}}{\mathrm{d}t} \tag{27}$$

where

$$a_g=\alpha_g\left(1-\rho_g\right) \tag{28}$$

$$a_w=\alpha_w\tau_g\left(1-\rho_g\right)\left(1-\rho_w\right) \tag{29}$$

$$a_{col}=\alpha_{col}\tau_g\tau_w\left(1-\rho_g\right)\left(1-\rho_w\right)\left(1-\rho_{\mathrm{col}}\right) \tag{30}$$

The unknowns in this set of equations are glass cover, brine and collector plate temperatures $\left(T_g, T_w, \& \ T_b\right)$. The equations are presented as the difference between the energy inputs and outputs equal the change of energy (accumulation or loss) in time.

Let us discuss briefly this set of equations. We consider that there are no sources of energy (chemical, electrical fissile, etc.) in any component of the still. The energy flux produced by the diffusion process is neglected. If all heat transfer coefficients, h, are temperature independent, then these differential equations are offirst-order and linear. Thus, analytical solutions can be found although the expressions can be quite complicated due to the time-dependence of solar irradiation, G. Notice that these equations would be nonlinear if the expression for heat radiation was used explicitly due to T^4. In this case, Equations (10) and (19) are solved for certain typical still parameters and their numerical values embedded into Equations (25), (26) and (27), so that the set can be solved easily.

These equations described lumped-parameter systems where no gradients are taken into account. Glass cover, brine and collector have properties independent of any coordinate direction. In this sense, they are analogous to the unitary operations in chemical and process engineering. These equations, modified dozens of times by researchers, have become the horse work in solar still modeling, providing accurate and useful information about its thermal behavior along a day.

Solar still studies have been, generally speaking, concerned aboutthe energy characterization of the device, putting aside the air-water vapor mixture flow inside the still. Experimentally [39] [41] [42] have discussed the flow structure and its time-dependence on solar stills in a laboratory.

4.3. Exergy Balance

Exergy is the maximum amount of work that can be extracted from a system when it goes from a certain thermodynamic state to a reference state [50] . The reference state is often modeled as the environmental state, although, depending on the type of exergy, some other references have been proposed ([51] -[53]). Exergy describes the quality of energy instead of its quantity for a given system or a flux; it quantifies how much energy goes directly into work. Simply put, exergy is the useful part of energy.

The term exergy was coined [54] by Zoran Rant in 1956 and its use has spread widely, especially in Europe. However, the concept can be traced back to Gibbs, namely Gibbs' available work. A detailed history of the exergy concept can be found in [55] .

The exergy balances refer to the account of exergy inputs, losses and the exergy destruction in the various components of the still. In mathematical form a general expression of this balance is:

$$B_{\text{out}} = B_{\text{in}} + B_{\text{source}} - \frac{\mathrm{d}B}{\mathrm{d}t} - I \tag{31}$$

It can be seen that the right-hand side of Equation (31) is formed by input exergy flux, B_{in}, the exergy from an energy source, B_{source}, (chemical, fissile, electrical, fusion, etc.), the local exergy change in time, $\frac{\mathrm{d}B}{\mathrm{d}t}$, and the exergy destruction by irreversibilities, I . Now, the exergy output may be a useful flux of exergy (work, heating, cooling, etc.) or simply exergy leaving the system with no useful consequence. However it may be, this form of exergy balance (Equation (31)) shows clearly the hindrances for having high values of exergy output. High thermal exergetic inertia and irreversibilities decrease exergy output.

The exergy balances for glass cover, brine and collector are, respectively:

$$\underbrace{B_{ev} + B_{cw} + B_{rw} + a_g B_S}_{\text{in}} - \underbrace{\left(B_{ca} + B_{ra}\right)}_{\text{out}} - \underbrace{I_g}_{\text{lost}} = \frac{\mathrm{d}B_g}{\mathrm{d}t} \tag{32}$$

$$\underbrace{a_w B_S + B_w}_{\text{in}} - \underbrace{\left(B_{ev} + B_{cw} + B_{rw}\right)}_{\text{out}} - \underbrace{I_w}_{\text{lost}} = \frac{\mathrm{d}B_w}{\mathrm{d}t} \tag{33}$$

$$\underbrace{a_{col}B_S}_{\text{in}} - \underbrace{\left(B_{colw} + B_{ins}\right)}_{\text{out}} - \underbrace{I_{col}}_{\text{lost}} = \frac{\mathrm{d}B_{col}}{\mathrm{d}t} \tag{34}$$

No mystery here. All energy terms in Equations (25)-(27) have its correspondent exergy plus the irreversibility that occurs in that component of the still. However, there is a crucial difference between energy and exergy balances. Exergy is a quantity that it is only conserved in a reversible process. When it is not conserved, irreversibility is produced. In an energy balance, no such concept is found because exergy is a concept derived from the 1st (energy) and the 2nd (entropy) law altogether. In this sense, thermodynamically speaking, "balance" is not the most suitable word to describe what happens to the exergy of a system in a process, nonetheless it is used here to describe more clearly in a mathematical sense.

The source for irreversibility can be friction, electrical resistance, accelerations, turbulence in a fluid, a temperature or a concentration difference among several others [56] [57] . Irreversibility is an amount of energy that can never be used to perform work. Eventually, all irreversibilities are dissipated as heat at ambient temperature.

In Equations (32)-(34), not all exergy fluxes have the same expression. For heat radiation, the exergy is expressed as [58] :

$$B = G\left(1 + \frac{1}{3}\left(\frac{T_a}{T}\right)^4 - \frac{4T_a}{3T}\right) \tag{34}$$

For all other heat transfer mechanisms, the exergy is calculated as [50] :

$$B = q\left(1 - \frac{T_a}{T}\right) \tag{34}$$

Considering that every component of the still is an incompressible substance (the air-water vapor mixture is not included in the balances), the change of exergy in time can have be expressed as:

$$\frac{\mathrm{d}B}{\mathrm{d}t} = \left(1 - \frac{T_a}{T}\right)mc\frac{\mathrm{d}T}{\mathrm{d}t} \tag{34}$$

or, when considering the possibility of a variable ambient temperature, as [59] :

$$\frac{dB}{dt} = mc\left[\left(1-\frac{T_a}{T}\right)\frac{dT}{dt} - \ln\left(\frac{T}{T_{ref}}\right)\frac{dT_a}{dt}\right] \tag{34}$$

Useful exergy is that part of the exergy used to heat the brine and produce distillation. Distillate production can be directly related to the solar exergy input. However, it is difficult to interpret exergy in terms that are not useful work (energy flow) because the amount of exergy from the distillate is not relevant. The fundamental expression for exergy has been established in most modern thermodynamics treatises, essentially as the difference between reversible work and lost work, which corresponds to a power maximization point of view. Thus, a high exergy air-water vapor mixture inside the still will produce more work than a low exergy mixture. The largest irreversibility in a solar still (and in any solar thermal application) occurs at the collector because a significant fraction of the total solar radiation absorbed in this component is poorly employed [59] . The high exergy losses causing the low exergyefficiencies are made up of avoidable and unavoidable contributions. The avoidable contributions can be reduced by improving the corresponding physical components of the system.

It is the opinion of the authors that second law analysis of solar-driven technologies, whether they are based on exergy, entropy generation, emergy, thermoeconomics or irreversible thermodynamics should complete any conventional thermodynamic analysis.

5. CONCLUSIONS

An extended overview of solar distillation from the thermodynamic point of view is presented in this work. Solar distillation has been known and applied for several hundreds of years; however, the understanding of physical phenomena behind it has been developed just some decades. A comprehensive and structured discussion is, therefore, necessary to a better understanding of the thermodynamic basis of these devices. Firstly, a concise history of solar distillation engineering is presented, from practice to theory. Relevant aspects of radiation are treated due to the importance of this parameter in solar distillation. The description of typical parameters in a solar still is useful for the identification of trade-offs in the performance of the device when one of them is varying. A considerable amount of literature is presented for each parameter's subsection.

Internal and external heat transfer phenomena in a solar still are clearly specified, including their definitions, mathematical modeling and a brief discussion of studies in the field for each one of them. One of the most important topics in solar distillation is thermodynamic modeling and the result is the energy balance. The authors have widened the scope of this study by the addition of second law concepts, such as exergy and irreversibilities. Intentionally, no mention is made to the several types of solar stills in the field and even more modifications are made to each of these types in order to increase distillation. In general, with some variations, every solar still performs based on the same phenomena is discussed here.

REFERENCES

1. Chaibi, M.T. and El-Nashar, A.M. (2009) Solar Thermal Processes. In: Cipollina, A., Micale, G. and Rizzuti, L., Eds., Seawater Desalination: Conventional and Renewable Energy Processes, Springer, London, 131-163.
2. Arellano, N. (2011) La planta solar de desalación de agua de Las Salinas (1872). Literatura y memoria de una experiencia pionera. Quaderns d'Història de l'Enginyeria, 12, 229-251.
3. Nebbia, G. (2005) Early Work on Solar Distillation in Italy, 1953-1970. In: Goswami, D.Y., Vijayaraghavan, S. and Campbell-Howe, R., Eds., Proceedings of the Solar World Congress 2005, American Solar Energy Society, Boulder, 2709-2713.
4. Kalogirou, S. (2005) Seawater Desalination Using Renewable Energy Sources. Progress in Energy and Combustion Science, 31, 242-281.
5. Maurel, A.(1981) La desalinización del agua de mar. Mundo Científico, 1, 296-305.
6. Malik, M.A.S., Tiwari, G.N., Kumar, A. and Sodha, M.S. (1982) Solar Distillation: A Practical Study of a Wide Range of Stills and Their Optimum Design, Construction and Performance. Pergamon Press, Oxford.
7. Telkes, M. (1953) Fresh Water from Sea Water by Solar Distillation. Industrial & Engineering Chemistry, 45, 1108-1114.
8. Dunkle, R.V. (1961) Solar Water Distilation: The Roof Type Still and a Multiple Effect Diffusion Still. Proceedings of International Heat Transfer Conference,University of Colorado, Boulder, Colorado, Part V, 895-902.

9. Morse, R.N. and Read, W.R. (1968) A Rational Basis for the Engineering Development of a Solar Still. International Journal of Solar Energy, 12, 5-17.

10. Cooper, P.I. (1969) Digital Simulation of Transient Solar Still Processes. International Journal of Solar Energy, 12, 313-331.

11. Ali, H.M. (1993) Effect of Forced Convection inside the Solar Still on Heat and Mass Transfer Coefficients. Energy Conversion and Management, 34, 73-79.

12. Singh, A.K. and Tiwari, G.N. (1992) Performance Study of Double Effect Distillation in a Multiwick Solar Still. Energy Conversion Management, 33, 175-181.

13. Sharma, V.B. and Mullick, S.C. (1993) Calculation of Hourly Output of a Solar Still. ASME Journal of Solar Energy Engineering, 118, 1-6.

14. Shawaqfeh, A.T. and Farid, M.M. (1995) New Development in the Theory of Heat and Mass Transfer in Solar Stills. Solar Energy, 55, 527-535.

15. Mowla, D. and Karimi, G. (1995) Mathematical Modelling of Solar Stills in Iran. Solar Energy Journal, 55, 389-393.

16. Tiwari, G.N., Singh, H.N. and Tripathi, R. (2003) Present Status of Solar Distillation. Solar Energy, 75, 367-373.

17. World Data Bank (2011) World Development Indicators. United States.

18. World Energy Council (2013) World Energy Resources: Solar.

19. Modest, F.M. (2003) Radiative Heat Transfer. 2nd Edition, Academic Press, Boston.

20. Gearhart, C.A. (2002) Planck, the Quantum and the Historians. Physics in Perspective, 4, 170-215.

21. Duffie, J.A. and Beckman, W.A. (2006) Solar Engineering of Thermal Processes. 3rd Edition, Wiley, Hoboken.

22. Garrison, J.D. and Roeder, S.B.W. (1999) Environmental Measurement. In: Webster, J.G., Ed., Measurement, Instrumentation and Sensors Handbook, CRC Press, Boca Raton, pp. 73.1, 1-64.

23. Anonymous (2002) Units and Symbols in Solar Energy. Solar Energy, 73, III-V.

24. Aboul-Enein, S., El-Sebaii, A.A. and El-Bialy, E. (1998) Investigation of a Single-Basin Solar Still with Deep Basins. Renewable Energy,

14, 299-305. McAdams, W.H. (1954) Heat Transmission. McGraw-Hill, New York.

25. El-Sebaii, A.A. (2000) Effect of Wind Speed on Some Designs of Solar Stills. Energy Conversion and Management, 41, 523-538.

26. Kumar, S. and Tiwari, G.N. (1996) Performance Evaluation of an Active Solar Distillation System. Energy, 21, 805-808.

27. Tiwari, G.N., Kumar, S., Sharma, P.B. and Khan, M.E. (1996) Instantaneous Thermal Efficiency of an Active Solar Still. Applied Thermal Energy, 16, 189-192.

28. Yadav, Y.P. and Prasad, A.S. (1995) Performance Analysis of a High Temperature Solar Distillation System. Energy Conversion and Management, 36, 365-374.

29. Tripathi, R. and Tiwari, G.N. (2005) Effect of Water Depth on Internal Heat and Mass Transfer for Active Solar Distillation. Desalination, 173, 187-200.

30. Tayel, S.A., El-Nakib, A.A., El-Meseery, A.A. and Badr, M.M. (2009) Solar Energy Utilization in Water Distillation. Misr Journal of Agricultural Engineering, 26, 428-452.

31. Badran, O.O. (2007) Experimental Study of the Enhancement Parameters on a Single Slope Solar Still Productivity. Desalination, 209, 136-143.

32. Keith, F. and Kreider, J.F. (1978) Principles of Solar Engineering. Hemisphere, Washington DC.

33. Rahbar, N. and Esfahani, J.A. (2013) Productivity Estimation of a Single-Slope Solar Still: Theoretical and Numerical Analysis. Energy, 49, 289-297.

34. Khalifa, A.J.N. (2011) On the Effect of Cover Tilt Angle of the Simple Solar Still on Its Productivity in Different Seasons and Latitudes. Energy Conversion and Management, 52, 431-436. Baum, V.A. and Bairamov, R. (1964) Heat and Mass Transfer Processes in Solar Stills of Hotbox Type. International Journal of Solar Energy, 8, 78-82.

35. Porta-Gándara, M.A., Chargoy, N. and Fernandez-Zayas, J.L. (1997) Extreme Operating Conditions in Shallow Solar Stills. Solar Energy, 61, 465-476.

36. Tiwari, G.N., Emran, K.M. and Goyal, R.K. (1998) Experimental Study of Evaporation in Distillation. Desalination, 115, 121-128.

37. Porta-Gándara, M.A., Rubio, E. and Fernandez J.L. (1998) Visualization of Natural Convection inside Shallow Solar Stills. Experiments in Fluids, 25, 369-370.

38. Zheng, H.F., Zhang, X.Y., Zhang, J. and Wu, Y.Y. (2002) A Group of Improved Heat and Mass Transfer Correlations in Solar Stills. Energy Conversion and Management, 43, 2469-2478.

39. Rubio-Cerda, E., Porta-Gándara, M.A. and Fernandez, J.L. (2000) Cavity Geometry Influence on Mass Flow Rate for Single and Double Slope Solar Stills. Applied Thermal Engineering, 20, 1105-1111.

40. Porta-Gándara, M.A., Cervantes, J.G. and Solorio, F.J. (2004) Periodic Enclosed Natural Convection in a Laboratory Solar Still. Experiments in Fluids, 37, 483-487.

41. Bird, B.R., Edward, W.E. and Lightfoot, E.N. (2007) Transport Phenomena. John Wiley & Sons, New York.

42. Lienhard IV, J.H. and Lienhard V, J.H. (2003) A Heat Transfer Textbook. Phlogiston Press, Cambridge.

43. Ahsan, A. and Fukuhara, T. (2009) Condensation Mass Transfer in Unsaturated Humid Air inside Tubular Solar Still. Annual Journal of Hydraulic Engineering, 53, 97-102.

44. Pong, L. and Moses, G.A. (1986) Vapor Condensation in the Presence of a Noncondensable Gas. Physics of Fluids, 29, 1796-1804.

45. Caruso, G., Di Maio, D.V. and Naviglio, A. (2013) Condensation Heat Transfer Coefficient with Noncondensable Gases inside Near Horizontal Tubes. Desalination, 309, 247-253.

46. Cussler, E.L. (2007) Diffusion, Mass Transfer in Fluid Systems. Cambridge University Press, New York.

47. Watmuff, J.H., Charters, W.W.S. and Proctor, D. (1977) Solar and Wind Induced External Coefficients for Solar Collectors. Revue Internationale d'Heliotechnique, 2, 56.

48. Bejan, A. (1998) Advanced Engineering Thermodynamics. Wiley, New York.

49. Szargut, J., Morris, D.R. and Steward, F.R. (1988) Exergy Analysis of Thermal, Chemical & Metallurgical Processes. Hemisphere Publishing Corporation, New York.

50. Serova, E.N. and Brodianski, V.M. (2004) The Concept "Environment" in Exergy Analysis: Some Special Cases. Energy, 29, 2397-2401.

51. Pons, M. (2009) On the Reference State for Exergy When Ambient Temperature Fluctuates. International Journal of Thermophysics, 12, 113-121.

52. Bosnjakovic, F. (1965) Technical Thermodynamics. Holt, New York.

53. Sciubba, E. and Wall, G. (2007) A Brief Commented History of Exergy from the Beginnings to 2004. International Journal of Thermophysics, 10, 1-26.

54. Kotas, T.J. (1994) The Exergy Method of Thermal Plant Analysis. Butterworths, London.

55. Moran, J.M. (1989) Availability Analysis: A Guide to Efficient Energy Use. ASME Press, New York.

56. Petela, R. (1964) Exergy of Radiation. Journal of Heat Transfer, 86, 187-192.

57. Torchia-Nu?ez, J.C., Porta-Gandara, M.A. and Cervantes-de Gortari, J.G. (2008) Exergy Analysis of a Passive Solar Still. Renewable Energy, 33, 608-616.

9

CHAPTER

Circular Thermodynamics of Organisms and Sustainable Systems

Mae-Wan Ho

Institute of Science in Society, 29 Tytherton Road, London, N19 4PZ, UK

ABSTRACT

A circular thermodynamics of organisms and sustainable systems is presented based on dynamic closures in nested space-time domains that enable the system to approach the ideal of zero entropy production simultaneously at equilibrium and far from equilibrium conditions.

Keywords: circular economy of nature; dynamic closure; space-time structure; zero-entropy model; quantum coherence; minimum entropy production; equilibrium and non-equilibrium

1. INTRODUCTION

The circular thermodynamics of organisms and sustainable systems has been developed in three successive editions of *The Rainbow and the Worm, The Physics of Organisms* [1] and elsewhere (especially [2,3]). The basic argument of this review article goes as follows.

1) A comparison of information processing in current supercomputers and the human brain reveals the enormous density and efficiency of

reactions in the living system that most likely involves quantum coherence.

2) The organism as a whole captures, stores, and mobilizes energy in a perfectly coordinated and super-efficient way, while maintaining its organization (homeostasis); these key features require a special thermodynamic explanation that dovetails with quantum coherence.

3) Energy is mobilized predominantly in cycles manifesting as biological rhythms with cycling times ranging from split seconds to minutes, hours, days, and years that are nevertheless coupled together. Cycles confer both dynamic stability and autonomy to the system. They also enable the activities to be coupled together, so that energy yielding processes can transfer energy directly to those requiring energy. Thermodynamically, no net entropy is generated in the case of perfect cycles; hence, the system can maintain its organization.

4) The organism has a nested dynamical structure, with activities spanning a wide range of characteristic space-time scales. The nested space-time structure partitions the organism into a hierarchy of systems within systems *defined by the extent of equilibration of thermal energies*. Thus, energies thermalized or equilibrated within a smaller space-time will still be out of equilibrium in the larger system encompassing the first, and hence capable of doing work. It is that which enables the organism to *simultaneously* achieve the most efficient equilibrium and non-equilibrium energy transfer.

5) The organism's nested dynamical structure also optimizes the kinetics of energy mobilization. For example, biochemical reactions depend strictly on local concentrations of reactants, which are very high, as their extent of equilibration is typically at nanometer dimensions.

6) The space-time differentiation of the living system is directly proportional to its energy storage capacity or energy residence time, which is also proportional to the material residence time; in ecosystem/sustainable system terms, these translate to biodiversity, biomass, or productivity.

7) In the ideal—approached most closely by the healthy mature organism and the healthy mature ecosystem—an overall internal conservation of energy and compensation of entropy ($\Sigma\Delta S = 0$) is achieved. In this state of balance, the system organization is maintained and dissipation minimized; *i.e.*, the entropy exported to the environment also approaches zero, $\Sigma\Delta S \geq 0$. Internal entropy compensation (and energy conservation) implies that there is free variation in microscopic states

within the macroscopic system; *i.e., the internal microscopic detailed balance at every point of classical steady state theory is violated.*

8) The above considerations give rise to the prediction that a sustainable system maximizes cyclic, non-dissipative flows while minimizing dissipative flows, *i.e.*, it tends towards minimum entropy production.

9) Some applications of the theory are presented to show how it disposes of major myths such as: sustainable systems mean no growth at all; a constant carrying capacity defines a plot of land; and the equation of energy with money.

2. THE COMPUTER *VERSUS* THE BRAIN

Scientists are attempting to put the human brain into a supercomputer. They propose to map all of the ~100 billion neurons in the human brain and their ~100 trillion connections (synapses) and then simulate it with the fastest supercomputer available.

There are fundamental reasons why the much touted European Human Brain Project and its U.S. counterpart, Brain Activity Map, will not achieve their goal. The precise connections between neurons in the brain vary from one individual to another, even between identical twins; memory is not stored in synapses, and above all, brains/minds do not work like a classical computer [4]; it could act more like a still theoretical quantum computer, though I have serious doubts as to whether a quantum computer can ever be practically implemented [5]. There is also a technical bottleneck most relevant to the thermodynamic argument presented here; the supercomputer cannot cope with the waste heat generated by the processing operations required to simulate the brain [6]. This is where the brain excels.

The human brain, about 1.2 L in volume, has an estimated memory of 3.5×10^{15} bytes, operates at a speed of 2.2 petaflops (10^{15} operations per second), and consumes 20 W in power. In comparison, the world's fastest supercomputer in June 2011 was Fujitsu's K computer the size of a small warehouse, has a memory of 30×10^{15} bytes, operates at a speed of 8.2 petaflops, and consumes 12.6 MW [7]. A year later, K computer was ousted as the world's fastest computer by the American IBM Sequoia, a Blue Gene/Q supercomputer performing at 16.325 petaflops using 123% more Central Processing Unit (CPU) processors and consumes only 7.9 MW, 37% less than the K computer [8]. However, China's Tianhe-2

(Milkyway-2) has now overtaken the U.S. supercomputer by performing twice as fast, at 30.7 petaflops [9], and the race is continuing apace.

The vast amount of power consumed by current computers generates a great deal of heat that has to be carried away by bulky cooling systems that consume yet more power. The human brain is a clear winner by far in terms of size and power consumption. The functional density (flops/L) in current computers is 10^{10} compared to 10^{14} in the brain; current computers carries out ~10^{7} operations per Joule of energy consumed, while the brain does 10^{14} operations for the same amount of energy [6].

The human brain is 2% of the body's volume and consumes 20% of its total energy demand; but it is fantastically energy efficient. It can achieve 7 orders of magnitude more computing for each Joule of energy consumed, while also performing in parallel all the activities required for keeping the brain cells alive. That is one reason I have proposed that organisms are quantum coherent [1]. The other main reason is, in 1992, we discovered in my laboratory that all living organisms display dynamic liquid crystalline interference (rainbow) colors under the polarizing light microscope [10,11,12]. The fact that living moving organisms, with all their molecules churning round transforming energy could still display colors typical of liquid crystals is evidence that living organisms are coherent (organized) to a high degree, right down to the alignment *and motions*of the macromolecules—and associated living water [13] in their tissues and cells—and it is *coherent* energy that is being mobilized and transformed in organisms [1,14]. Specifically, organisms are powered by *quantum molecular machines* that transform and transfer energy at close to 100% efficiency, which is why the brain can do 10^{14} operations per second in processing information and, at a conservative guess, another 10^{11} metabolic reactions per second to keep the cells alive. Such density of function cannot be achieved classically, and requires quantum coherence of a high order (close to zero energy dissipation). Consequently, it also calls for a thermodynamics that dovetails with quantum coherence, and that is what I shall present here.

3. THERMODYNAMICS OF LIVING SYSTEMS

The thermodynamics of living systems was formulated over a period of some 16 years, most fully detailed in successive editions of *The Rainbow and the Worm, the Physics of Organisms* [1]. I shall recapitulate the main results and bring this work up to date.

The first thing to take note is that organisms do *not* make their living by heat transfer. Instead, they are isothermal systems (Morowitz [15]) dependent on the direct transfer of molecular energy, by proteins and other macromolecules acting as "molecular energy machines" (more specifically, *quantum* molecular energy machines). For isothermal processes, the change in Gibbs free energy ΔG (thermodynamic potential at constant temperature and pressure) is,

$$\Delta G = \Delta H - T\Delta S \tag{1}$$

where ΔH is the change in enthalpy (heat content), T is the temperature in deg K. and ΔS the change in entropy.

Thermodynamic efficiency requires that ΔS approaches 0 (least dissipation) and $\Delta H =$ 0; or ΔG approaches 0 via entropy-enthalpy compensation, *i.e.*, entropy and enthalpy changes cancelling each other out [1,16].

The organism as a whole keeps far away from thermodynamic equilibrium (in the classical sense), but how does it free itself from "all the entropy it cannot help producing while alive?" was the question that Schrödinger asked [17]). That is also my point of departure for the thermodynamics of living systems.

The pre-requisite for a system to keep away from thermodynamic equilibrium—the state of maximum entropy or death by another name—is to be capable of *capturing* energy and material from the environment to develop, grow and recreate itself from moment to moment during its life time, and also to reproduce and provide for future generations, all part and parcel of sustainability.

The organism has solved the problem of sustainability over billions of years of evolution. It has an obviously nested physical structure. Our body is enclosed and protected by a rather tough skin, but we can exchange energy and material with the outside, as we need to, we eat, breathe and excrete. Within the body, there are organs, tissues and cells, each with a certain degree of autonomy and closure. Within the cells, there are numerous intracellular compartments that operate more or less autonomously from the rest of the cell. And within each compartment, there are molecular complexes doing different things: transcribing genes, making proteins and extracting energy from our food, *etc.*, all working within confines of nanometre dimensions (nanospaces). More importantly, the activities in all those compartments, from the microscopic to the macroscopic are perfectly orchestrated, which is why the organism looks like a dynamic liquid crystal display, as explained earlier.

It can be questioned whether physical closure is necessary, at least as far as the sustainable system is concerned; more important than physical closure is *dynamic* closure, which enables the organism to *store* as much energy and material as possible, and to use the energy and material most efficiently, *i.e.*, with the *least* waste and dissipation.

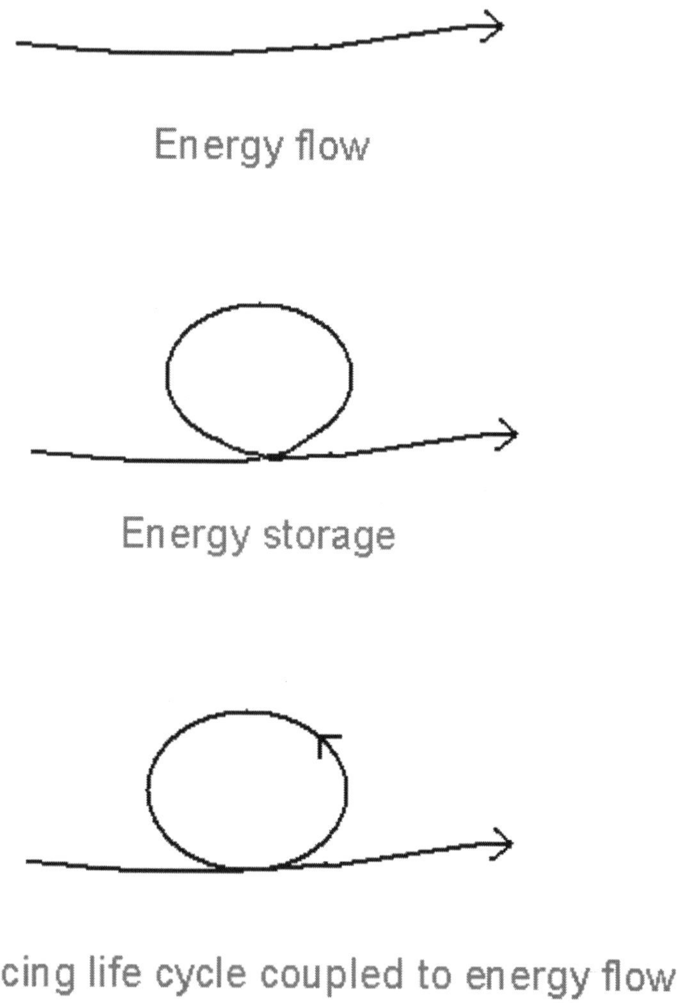

Figure 1. Energy flow, energy storage and the reproducing life-cycle [1].

The key to understanding the thermodynamics of the living system is not so much energy flow (stressed by many commentators, for example, Prigogine [18], Morowitz [15], and Ulanowicz [19]) as energy *capture and storage* under energy flow (Figure 1). Energy flow is of no consequence unless the energy is trapped and stored within the system, where it is mobilized to give a self-maintaining, self-reproducing life cycle coupled to the energy flow. (By energy, I include material flow, which enables the energy to be stored and mobilized.)

My approach diverges significantly from the framework established by earlier applications of thermodynamics to ecology as described in detail in [3]. For example, *stored* energy, as I shall define it more precisely, is distinct from *exergy* as widely used by ecologists, and also from *free energy* as defined by chemists and physicists (Equation (1)). It is stored energy being mobilized in a *non-classical* steady state (as will be made clear) that characterizes living organisms and sustainable systems.

4. CYCLES MAKE SENSE

The perfect coordination (organization) of the organism depends on how the captured energy is mobilized within the organism. It is actually mobilized in *cycles*, or more precisely, quasi-limit cycles, which can be regarded as dynamic boxes; and they come in all sizes, from the very fast to the very slow, from the global to the most local.

Cycles provide the dynamic closure that's absolutely necessary for life, much more so than physical closure.

Biologists have long puzzled over why biological activities are predominantly rhythmic or cyclic, and much effort has gone into identifying the center of control, and more recently to identifying master genes that control biological rhythms, to no avail.

The organism is full of cycles (as is the earth, for that matter), possibly because cycles make thermodynamic sense; Morowitz [15] has proven an important theorem that a flow of energy from a source to a sink in a system at steady state will lead to at least one cycle, so there is definitely a deeper reason why cycles tend to form under energy flow. (Intuitively, I suspect energy storage is important; it is like water forming eddies flowing through a pool, a consequence of the conservation of momentum.) Cycles mean returning repeatedly to the same states, and no entropy is generated in perfect cycles according to classical thermodynamics. In other words, the system as a whole remains organized and no waste heat is produced; free energy change is zero because the energy has not been degraded into thermal energy. Cycles give dynamic stability as well as autonomy to the organism; and this applies as well in ecosystems [19].

Cycles also enable the activities to be coupled, or linked together, so that those yielding energy can transfer the energy directly to those requiring energy, and *the direction can be reversed when the need arises*. This is implicit in Onsager's reciprocity relationship [20], which shows how symmetrical coupling of processes (through forces and flows) can arise naturally in a system under energy flow at equilibrium or near equilibrium, as pointed out since the first edition of *Rainbow Worm* [1]. These symmetrical, reciprocal relationships are most important for sustaining the system. Our metabolism is organized precisely in that way: closing cycles and linking up, with pathways that readily reverses the direction of energy and material flows.

Figure 2 is a diagram representing the nested cycles that span all space-time scales, the totality of which make up the life cycle of the organism (see 2nd and 3rd editions of [1]). I proposed that the life cycle has a self-similar fractal structure (simply because natural processes are known to be predominantly fractal [21], and in my view, fractal mathematics is the only tractable way to describe complex organic space-time). Therefore, if you magnify each cycle, you will see smaller cycles within, looking much the same as the whole. Fractal dynamics are the hallmarks of natural processes and are especially fit for the organization of living systems, as we shall see.

The complex nested dynamical space-time structure of the organism is the secret of sustainability. It maximizes the efficiency and rapidity of energy mobilization, and the degree of space-time differentiation is directly correlated with the amount of energy stored.

5. REDEFINING THE SECOND LAW FOR LIVING SYSTEMS

Physiologist Colin McClare [22] made an important contribution towards reformulating thermodynamics so that it can apply to living systems, especially to energy transfer between molecules, which is what happens in living systems. He proposed that in a system defined by some macroscopic parameter, such as temperature, θ, its energies can be separated into two categories: *stored* (coherent) energies that remain in a non-equilibrium state within a characteristic time, τ, and*thermal* (random) energies that exchange with each other and reach equilibrium (or equilibrate) in a time less than τ (Figure 3).

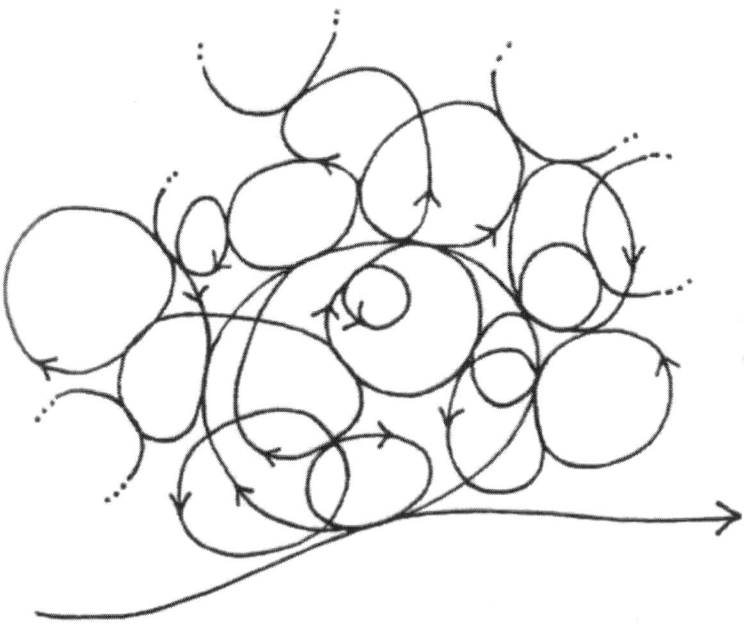

Figure 2. The life cycle of the organism consists of a self-similar fractal structure of cycles turning within cycles [1].

McClare introduced time structure into systems, with the very important consequence that there are now *two* ways to mobilize energy efficiently with entropy change approaching zero: very slowly with respect to τ, so it is reversible at every point; or very rapidly with respect to τ, so that the energy remains stored (in a coherent or non-degraded form) as it is mobilized.

For a process with characteristic timescale of 10^{-10} s, a millisecond is an eternity, so a "slow" process need not be very slow at all from our point of view to be energy efficient. Most enzyme reactions in the living system therefore could be occurring at or near thermodynamic equilibrium.

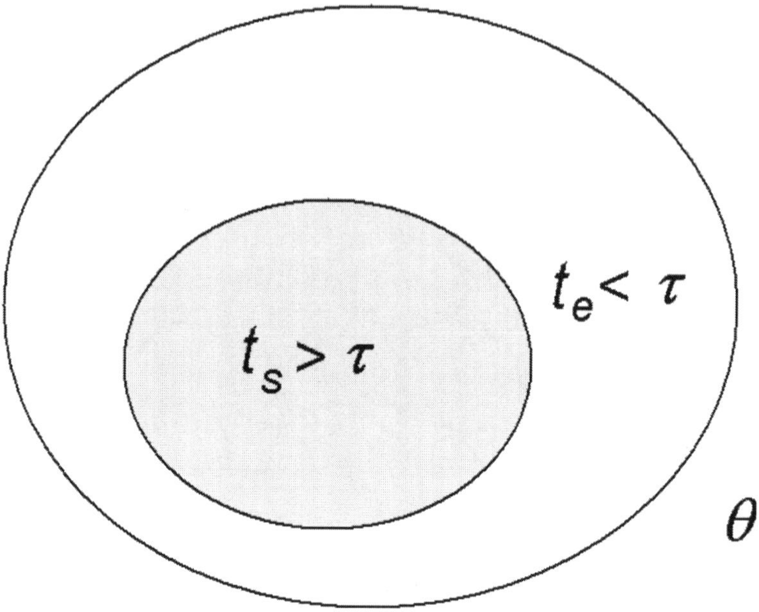

Figure 3. Stored *vs.* thermal energy [1].

On the other hand, resonant energy transfer is an example of a very fast process occurring in $<10^{-14}$ s, so the energy remains stored as it is transferred. Resonant energy transfer is also very important for living systems. Resonance interactions coordinate reactions in different parts of the cell and the organism. Resonating molecules attract one another, and there is indeed evidence that proteins, nucleic acids and other molecules find one another by resonating to the same electromagnetic frequencies [23], and not by random collision in aqueous solution as represented in conventional biochemistry and enzymology (see [1,13]). For one thing, the cell is an extremely crowded and organized environment for molecules.

McClare [22] proposed that, "Useful work is only done by a molecular system when one form of stored energy is converted into another". In other words, thermalized energies cannot be used to do work, and thermalized energy cannot be converted into stored energy. This proposal raised obvious objections, as critics pointed out, automobiles *do* run on thermalized energy from burning petrol, so the proposal is not correct for automobiles and cannot be correct for molecular systems, if molecular systems do not violate the Second Law of thermodynamics.

McClare's proposal was incomplete, and I completed his proposal as follows [1,16]: "Useful work is only done by a molecular system when one form of stored energy is converted into another *in the same system*." The additional phrase "*in the same system*" effectively defines a "system" by *the extent to which thermal energies equilibrate within a characteristic space-time*. This also generalizes McClare's characteristic time of processes to characteristic space-time of processes.

In the case of the automobile and other similar contraptions, the hot gases expand against a constraint, the piston, which, in taking up the thermalized energy, does work against the system *external* to the combustion chamber.

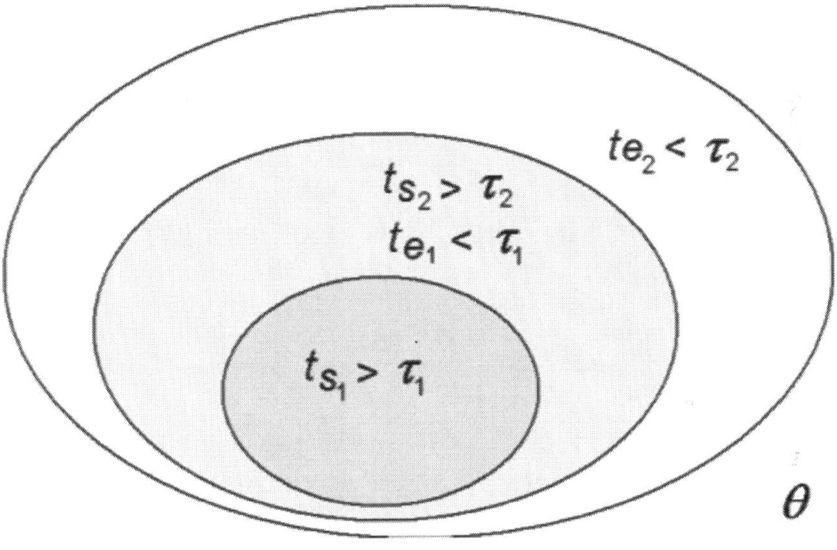

Figure 4. A nested hierarchy of space-times in which equilibrium and non-equilibrium canco-exist [1].

This definition of a system is crucial for the nested space-time structure of the organism. As the organism is partitioned into a hierarchy of systems within systems within systems defined by equilibration space-times, energies thermalized or equilibrated within a smaller space-time (system) will still be out of equilibrium in the larger system encompassing the first (Figure 4). So, even though the organism as a whole is far from thermodynamic equilibrium, its space-time differentiation nevertheless allows for a hierarchy of local near-equilibrium regimes to be maintained within.

Stored energy, like exergy and free energy, refers to energy available for doing useful work, it is coherent energy. But stored energy is explicitly defined with respect to a characteristic space-time, and is hence a real property of systems rather than a pseudo-property (see [3]).

The nested space-time structure in organisms optimizes thermodynamic efficiency by allowing the organism to simultaneously exploit equilibrium (very slow) and non-equilibrium (very fast) energy transfers with minimum dissipation, always with reference to the characteristic times of the processes involved as described above. It also optimizes the rapidity of energy mobilization. Biochemical reactions depend strictly on local concentrations of reactants, which could be enormously high, depending on their extent of equilibration. Cell biologists are beginning to take seriously the view that the cell approaches the solid-state, or more accurately, a liquid crystalline state, where nothing is freely diffusible, and even the cell water is organized into polarized multi-layers [1,13,14] as first proposed by Gilbert Ling [24]. Typically, molecular interactions occur within nanospaces both inside and outside cells. The nanospace defines the local system for molecular energy machines in the extent of equilibration (Figure 3, Figure 4).

Another point to note is that the greater the space-time differentiation, the more the amount of coherent energy effectively stored within the system. Because the activities are all coupled together, the energy residence time depends on how many activities there are within the system.

Finally, I conjectured that the fractal space-time structure in the organism allows activities at all scales to remain distinct and independent, and yet poised for energy exchange with one another. In other words, energies in different space-time domains are separately mobilized, and nevertheless able to spread from any point to the entire system, or conversely to converge from all over the system to any point whenever and wherever required [1,2,25]. However, it was only in 2009 that I suddenly got an inkling of why that might be the case.

As we were about to watch Simon McBurney's *A Disappearing Number* performed by the company Complicité at the National Theatre in London, a beautiful play (as it turned out) created around the Indian mathematical sensation Srinivasa Ramanujan, I asked Peter Saunders, my husband and mathematical adviser for more than 35 years, if all irrational numbers were arbitrarily close to rational numbers, and he said yes. My guess was that living fractals are associated with irrational numbers such as the golden mean, and cannot enter into resonance; which is why the activities can remain distinct. However, the fractals are also close to harmonics that do resonate, so phase coupling and energy transfer through resonance is readily achieved by sliding from fractals to harmonics. Sliding from fractals to harmonics may coincide with the threshold for percolation in which local connects with global. Panchelyuga and Panchelyuga [26] recently considered resonance and fractals and the distribution of rational *versus* irrational numbers on the real numbers set and their relevance for physical systems, though without relating to the hypothesis presented here.

Cell biologists have independently proposed that the cytoplasm can be described as a percolation cluster or a random fractal [27]. A percolation cluster is an ensemble of sites that become globally connected beyond a percolation threshold (see Chapter 18 of [13] for more details).

There is also abundant evidence that fractal dynamics characterize the healthy heart rhythm, which reflects the constant intercommunication between the heart and all other parts of the body [28]. Real time monitoring shows how the heart rhythm can change abruptly, and how positive emotions such as love and appreciation can make the heart beat in synchrony with the pulse and respiratory rhythms, possibly through sliding from fractals to harmonic resonance on a macroscopic scale [29].

6. THE "ZERO-ENTROPY" IDEAL

The healthy mature organism as well as the healthy mature ecosystem as defined by Odum [30] approaches an ideal dynamic balance of zero-entropy production (Figure 5). The simple equation $\Sigma\Delta S = 0$ inside the cycle, says there is an overall internal conservation of energy and compensation of entropy so that the system organization is maintained and dissipation reduced to zero—while the necessary dissipation exported to the outside, *is also minimized,* $\Sigma\Delta S \geq 0$. Readers will note that this zero-entropy ideal is also the ideal quantum coherent state [1]. I emphasize *ideal,* because the ideal system will never age and never die, whereas real organisms and ecosystems slowly age and eventually die, basically because real organisms interact with the environment and decohere before re-establishing coherence as an asymptotically stable state [1].

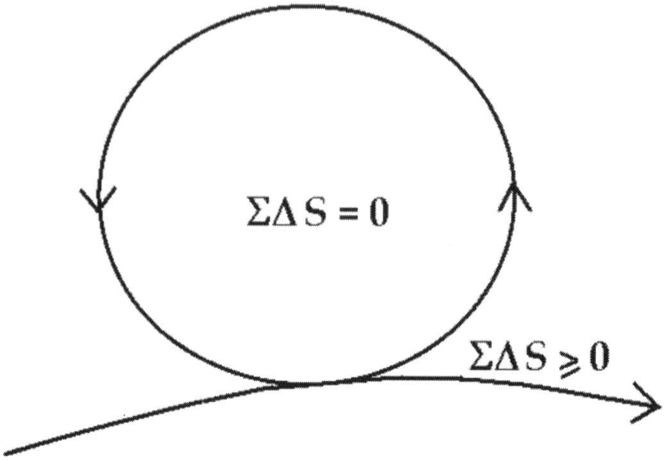

Figure 5. Zero-entropy model of the ideal organism and sustainable system [1].

Internal entropy compensation and energy conservation implies that positive entropy generated somewhere is compensated by negative entropy (or decrease in entropy) elsewhere within the organism *over a finite time.* This is just what Schrödinger [17] had tried to represent with the term "negentropy". *This is possible only if the internal microscopic detailed balance at every point of classical steady state theory is violated.*

Denbigh [31] defined the "steady state" as one in which "the macroscopic parameters such as temperature, pressure and composition, have time-independent values at every point of the system, despite the occurrence of a dissipative process". This is far too restriction for any real process, which always takes place in finite time and finite volume. Instead, I proposed [1] to define the living system in homeostasis as a "dynamic equilibrium in which the macroscopic parameters, such as temperature, pressure and composition have time-independent values despite the occurrence of dissipative processes". The omission of the phrase "at every point of the system" is important and significant.

Microscopic homogeneity is not necessary for the formulation of any thermodynamic state, as the thermodynamic parameters are macroscopic entities quite independent of the microscopic interpretations. Like the principle of microscopic reversibility, it is extraneous to the phenomenological laws of thermodynamics, as Denbigh himself had convincingly argued [31].

It is the organized space-time *heterogeneity* within the living system that allows for the necessary "free" variation of the microscopic states within the macroscopic thermodynamic constraints. Thus, stability criteria that apply to the system as a whole need not be satisfied; or stronger yet, *cannot* be satisfied in every individual space-time element for all times.

The tendency to conserve coherent energy and compensate for entropy production *within* the system will result in the minimum entropy being exported to the outside. Intuitively, one can see that if the system were maximally efficient, then it would also produce the least dissipation. *This is the basis for Prigogine's minimum entropy production principle* (see [18]), *which, in the present case, applies to conditions both near and far from thermodynamic equilibrium.*

From the outside, it might appear that the system is "maximally dissipative" in terms of having "degraded" the energy gradient most effectively [32,33,34]. But this misses the coherent energy *stored* non-degraded within the system, and stored energy is also embodied in matter within the system, *i.e.*, biomass.

7. SUSTAINABLE SYSTEMS AS ORGANISMS AND DIAGNOSTIC SIGNS OF SUSTAINABILITY

Based on the circular thermodynamics of organisms, I have suggested diagnostic criteria of sustainability or health that depend on the tendency of a sustainable system to maximize non-dissipative cyclic flows of energy and minimizing dissipative flows [25].

Maximizing non-dissipative cyclic flows will increase the following: energy storage capacity, which translates into carrying capacity or biomass; the number of cycles in the system; the efficiency of energy use; space-time differentiation, which translates into biodiversity; balanced flows of resources and energy; reciprocal coupling of processes. The minimization of dissipation will result in reducing entropy production (towards zero).

These diagnostic criteria are interlinked, so once one is identified, the others are likely to follow. Some support for these criteria is that they are similar to those Schneider and Kay [32] have identified for mature, established ecosystems [25]. Data collected for carbon-energy flows in two aquatic marsh ecosystems next to a large power-generating facility in the Crystal River in Florida showed that the "stressed" system, exposed to hot water coming out of the nuclear power station, which increased the temperature by 6 °C, captured 20% less energy, made 20% less efficient use of the energy captured, had 50% fewer cycles and 34% less biomass than the control.

Schneider and Kay [32] also drew attention to some interesting measurements made by Luvall and Holbo [35] with a NASA thermal infrared multispectral scanner from the air, which assess energy budgets of terrestrial landscapes. They found that the more developed the ecosystem, the *colder* its surface temperature. This is consistent with the maximization of energy storage capacity and the minimization of dissipation, as the system develops towards the zero-entropy ideal; rather than it being more effective at degrading the energy gradient, as interpreted by Schneider and Kay [32].

Another indication of the energy efficiency and potential increase in carrying capacity of sustainable systems is provided by a comparison of 25 rice cultivation system [25,36], of which 8 were pre-industrial in terms of low fossil fuel input (2–4%) and high labor input (35–78%), 10 were semi-industrial with moderate to high fossil fuel input (23–93%) and low to moderate labor input (4–46%) and seven were full industrial with 95% fossil fuel input and extremely low labor input of 0.04–0.2%. The total output per hectare (in gigajoules) in the pre-industrial fell into a low (2.4 to 9.9) and a high-output (149.3 to 166.9) subgroup, with the former one-twentieth to one-fifth of the full industrial average. However, the output of the high subgroup was two to three times the full-industrial systems. The yields of semi-industrial systems were more homogeneous, with an average of 51.75 GJ, while the yields of full-industrial systems, even more uniform, averaged 65.66 GJ.

When the ratio of total energetic output to total input was calculated, the pre-industrial low yielding systems ranged between 6.9 and 11.5, whiles figures for the high output system registered from 15.3 to 29.2. Semi-industrial systems gave ratios of 2.1 to 9.7, whereas the ratios of full-industrial systems were not much better than unity. These figures illustrate the law of diminishing returns: there seems to be a plateau of output per hectare around 70–80 GJ regardless of the total input, which is only exceeded in the three high-yielding pre-industrial systems of Yunnan, China. Intensifying energy input led to a drop in efficiency, particularly sharp as input approaches the output ceiling, which appeared to conform to the notion of a uniform carrying capacity. But this is highly misleading, as the carrying capacity depends on how the land is organized for production, in particular, soil fertility, as already suggested by the Yunnan rice production systems.

Another interesting study compared the whole of Danish agriculture from 1936 to 1990, during which, mechanization and chemical inputs in fertilizers and pesticides increased 20-fold, the average farm size more than doubled from 15.7 ha to 35.1 ha, and average crop yields for human food rose from 10 GJ/ha to 21.7 GJ/ha [37]. However, total energy input increased 8-fold over the period, from 2.6 GJ/ha to 21.1 GJ/ha; consequently, the energy output/input ratio fell from 3.9 to 1.0.

In both examples, the energy output/input ratio gives a good indication of sustainability and energy efficiency. Industrial monoculture not only pollutes the environment with its chemical and fossil fuel inputs, it harms the complex ecosystem of the soil that plays an essential role in nutrient recycling, water retention, as well as disease control, effectively putting a ceiling on agricultural productivity. In contrast, all forms of low-input sustainable agriculture depend on working with and within the circular economy of nature with maximum use of internal inputs, biological pest control, composting, green manure and farmyard manure (see [38] for a comprehensive report on the subject). There are more ways to approach the zero-entropy ideal of the organism and sustainable system. I shall begin with a conceptual example.

8. DREAM FARM 2

Some time ago, I proposed the concept of an integrated food and energy self-sufficient farm based on the circular thermodynamics of organisms; one of the most recent versions is given in Figure 6 [39]. Anaerobic digestion recycles plant, livestock, food and human wastes into biogas fuel and agricultural nutrients, and other green energies at small to micro-scale are also incorporated, such as solar, wind, microhydroelectric, as appropriate. Combined heat and power generation provides heating for buildings and thermoelectric waste heat harvesting and cooling could be incorporated for maximum energy recycling. Surplus electricity could be used for charging batteries for energy storage, as well as electric cars, tractors, computers, mobile phones *etc.* This mix of energies ensures a reliable supply, and can reduce energy use by at least 30% through exploiting 'waste' heat from power generation, use of heat pumps, and preventing energy loss in long distance distribution and transmission.

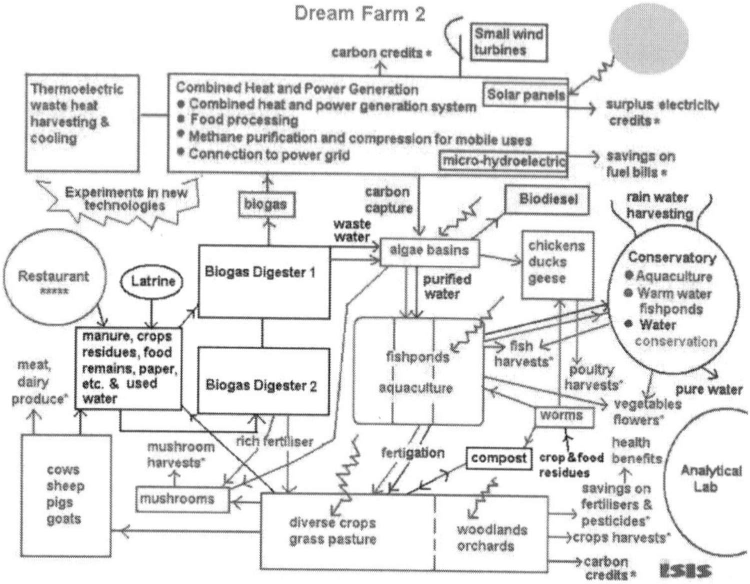

Figure 6. Dream Farm 2 based on the zero-entropy ideal of the organism [39].

The diagram is color-coded. Pink is for energy, green for agricultural produce, blue is for water conservation and flood control, black is waste in the ordinary sense of the word, which soon gets recycled and converted into food and energy resources. Purple is for education and research into new science and technologies. It is estimated to save 44% of energy consumption, only counting anaerobic digestion. With the addition the other renewables, there would be more than enough energy for the entire farm, (while also compensating substantially if not completely for carbon emissions) (see Table 1, estimated for China [39]).

Table 1. Green potential of Dream Farm 2 if adopted nationally for China [39].

	CO_2e savings	(%National)	Energy savings	(%National)
Organic agriculture	955.1 Mt	(12.69%)	2.608 EJ	(3.61%)
Anaerobic digestion	586.0 Mt	(7.79%)	7.470 EJ	(10.35%)
Energy savings local gen.	1,287.1 Mt	(17.10%)	21.660 EJ	(30.00%)
Total	**2,828.2 Mt**	**(37.58%)**	**31.738 EJ**	**(43.96%)**

Dream Farm 2 would be an excellent project for a university or a research institute because engineers, architects, scientists, artists, medical doctors, sociologists, economists and business can all work together across the disciplines to realize the circular, zero-entropy model in design, energy flow, architecture, marketing, *etc.*, while providing opportunities for education, research and innovation.

A more modest version of the scheme would be particularly suitable for providing off-grid energy (as well as sanitation in anaerobic digestion toilets) to isolated communities in developing countries [39]. Modern electronic lighting, mobile phones and other household equipment have low power requirements, and are ideally served by the local renewable energies available. Indeed, small anaerobic digesters, solar lanterns, and solar panels are already improving the lives of millions in isolated parts of China and Africa.

On the other hand, Dream Farm 2 can also be scaled up for a local community, say for visionary science, which includes everyone whom the research will benefit. There would be housing, childcare centers, restaurants, shops, theatre, concert hall, art gallery, a university, a high school, primary school, and a community bank.

9. PRODUCTIVITY, CARRYING CAPACITY & BIODIVERSITY

Dream Farm 2 illustrates how it *is* possible to have sustainable growth and development. Too many critics of the dominant paradigm think that the only alternative to unsustainable growth is to have no growth at all.

The minimum entropy exported to the environment is important, as the system depends on environmental input, hence, entropy exported to the environment will simply mean diminished environmental input. This can be made more explicit by enclosing the system within the immediate environment of the system as in Figure 7.

The ecological environment surrounding the system is now explicitly represented also as a zero-entropy cycle. Once again, that this is a fractal diagram, the environment surrounding the system is exporting to a larger ecological domain, and this kind of embedding can go on, ultimately to the entire earth. In addition, each cycle is of course made up of many smaller cycles within (see Figure 2) all working by reciprocity and cooperation. This principle is well illustrated in sustainable integrated farming.

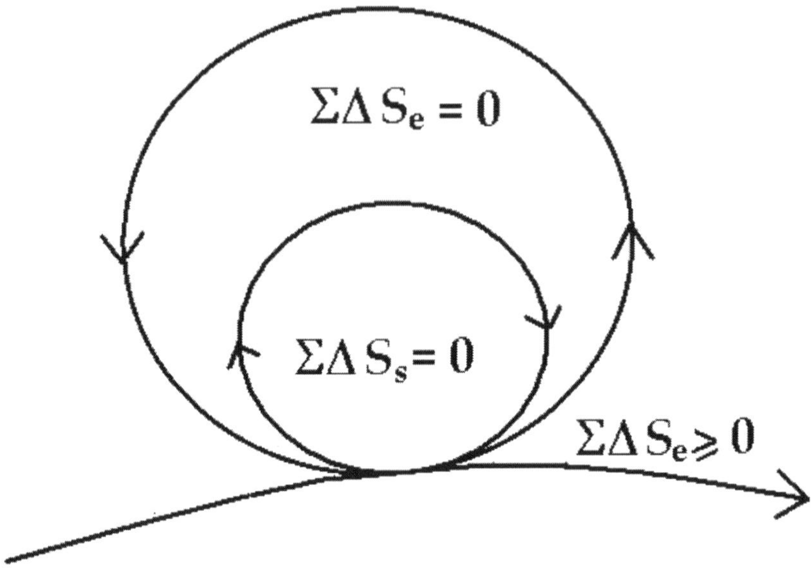

Figure 7. The coupled flows of system and ecological cycles in a sustainable system [1].

The minimum integrated farm has the farmer, livestock and crops (Figure 8). The farmer prepares the ground to sow the seeds for the crops to grow that feed the livestock and the farmer; the livestock returns manure to feed the crops. Very little is wasted or exported to the environment. In fact, a high proportion of the resources are recycled and kept inside the system. The system stores energy as well as material resources such as carbon. The extra carbon is sequestered in the soil as the soil improves, and in the standing biomass of crops and livestock.

More importantly, the farm can perpetuate itself like that quite successfully and sustainably, or it can grow by engaging more cycles, units of devolved autonomy that help one another do better. (In analogy with the organism, it will develop a more complex space-time differentiation, and grow bigger.)

In the old paradigm, organisms are predominantly seen to compete for resources and for space. However, we've got three space dimensions and the time dimension too. We've got space-time that can be filled up more thickly with life cycles of different sizes that occupy different space-times. So you can add fish, algae, poultry, worms, mushrooms, *etc.*, turning the 'waste' from one cycle to resource for another (Figure 9).

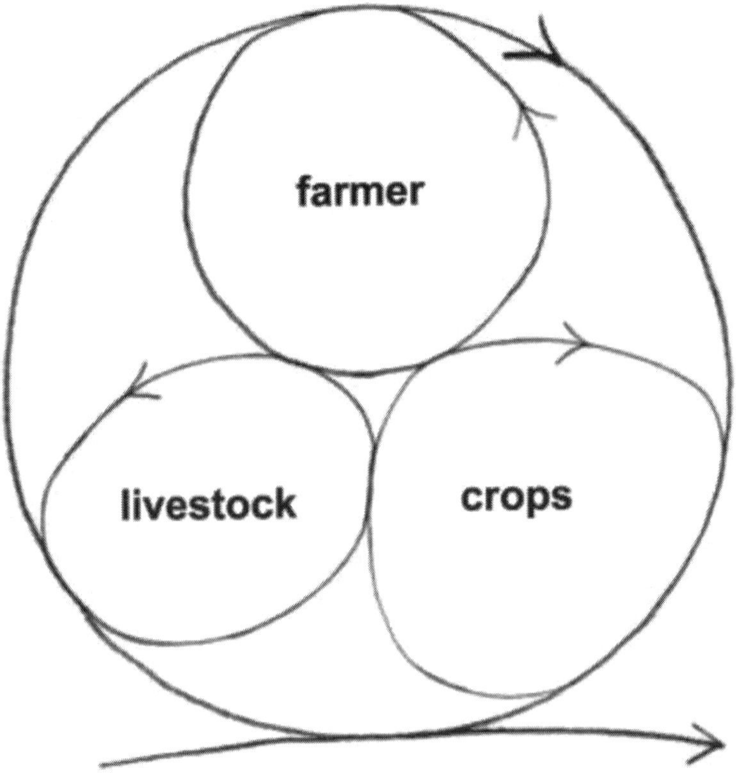

Figure 8. The minimum integrated sustainable farm [48].

The more lifecycles incorporated, the more energy and standing biomass are stored within the system, the greater the productivity of the farm and more farmers and farm workers supported. Chinese farmers have perfected the circular economy over the past two thousand years especially in the Pearl River Delta of southeast China [40]. Crops, livestock and silkworms are grown on the dykes, some overhanging the ponds, where up to five species of carp are fed on the wastes; the water and pond mud are used to fertilize the crops. Variants of such dyke-pond systems supported 17 people per ha in the 1980s, a carrying capacity at least ten times the average of industrial farming, and two to three times the world average. It disposes of another myth: that a constant carrying capacity exists for a given piece of land.

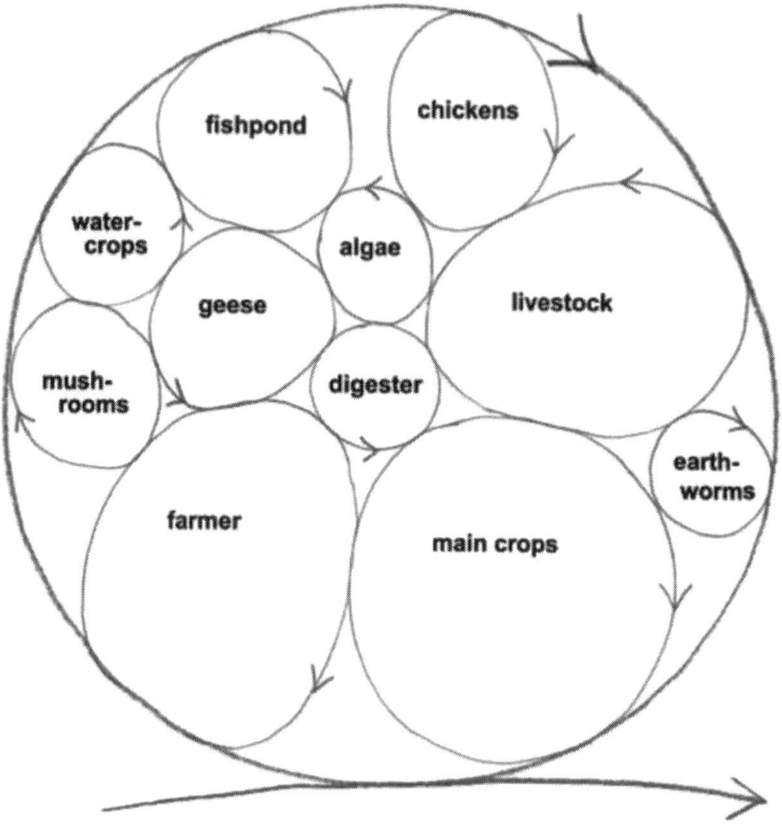

Figure 9. Sustainable system develops and grows by incorporating more life cycles within the system, the wastes from one cycle is resource for another [48].

Productivity and biodiversity *always* go together in a sustainable system, as generations of farmers have known, and recent academic researchers have rediscovered. The different life cycles are essentially holding the energy for the benefit of the whole system, and by way of reciprocity, *recycling* the energy within the system. Once it is recognized that coherent energy is *stored* within the system, it follows that energy *can* be recycled, contrary to the conventional wisdom that regards only materials as capable of being recycled.

That is exactly what happens in a naturally biodiverse ecosystem: organisms maximize the reciprocal, symbiotic relationships that ultimately benefit *all* the species; and biodiverse systems have been shown time and again to be more productive than monoculture cropping [41].

Physiologists have long observed that the sizes (body weights) B of animals scale allometrically with basal (resting) metabolic rate M:

$$B = B_0 M^{\alpha} \tag{2}$$

where B_0 and α are constants. This has remained a mystery until Geoffrey West, James Brown and Brian Enquist derived the scaling relationship from first principles, showing that it stems from the fractal structure of organisms, especially of circulatory and transport networks that are optimized for maximum exchange surface, minimum transport distances and times, and minimum dissipation [42]. This is certainly in line with the circular thermodynamics zero-entropy ideal model of organisms and sustainable systems proposed here [43].

10. CIRCULAR THERMODYNAMICS AND GREEN LIVING ECONOMY

Thermodynamics is the science of material and energy transformation, the circular thermodynamics of organisms is therefore none other than living economy: the transformation of energy and materials that enable organisms including human beings to survive and thrive.

It is of interest that "circular economy" is in Chinese mainstream discourse [44]; and is actually adopted as a national policy, albeit only in the manufacturing and service sectors; they had neglected to include agriculture [45], the primary production for all economies. Circular economy is also entering the mainstream in the world at large, as championed by Ellen MacArthur, who has started the "Circular Economy 100" for the business world [46], so it is all the more important to understand what circular economy is all about.

There is a world of difference between circular integrated farming and the linear input-out industrial agriculture. It is also the difference between zero-entropy circular economy that some of us aspire to, and the dominant economy of infinite competitive growth and maximum dissipation (entropy generation). I contrast the two in Figure 10.

The dominant linear input-output system grows relentlessly, swallowing up the earth's resources, laying waste to everything in its path, like a hurricane. There is no closed cycle to hold resources within, to build up stable organized social or ecological structures. That's the essence of our "boom and bust" economy.

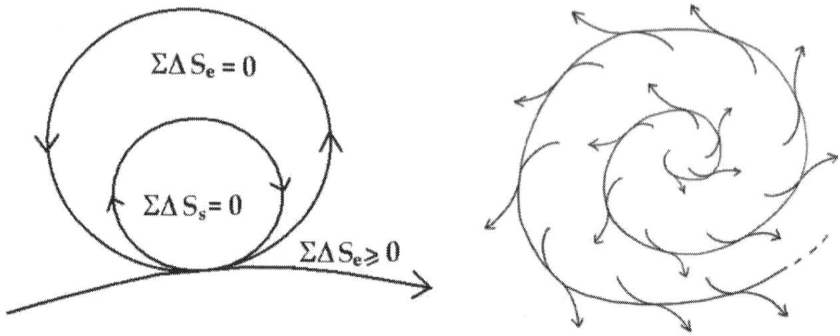

Figure 10. The zero-entropy circular economy patterned after and integrated with the circular economy of nature (**left**) *versus* the dominant economic model of infinite unsustainable growth that swallows up the earth's resources and exports massive amounts of wastes and entropy (**right**).

The zero entropy circular economy, on the other hand, is embedded within and integrated with the circular economy of the natural ecosystem. It builds up space-time structures within to store and mobilize renewable energy and materials through values added in various ways to the primary productivity from sunlight and photosynthesis.

As argued earlier, the fractal structure of the circular economy (see Figure 2) ensures that all space-times are bridged, and is optimized for capturing, storing, and mobilizing resources efficiently and rapidly while conferring local autonomy on all scales. It also maximizes diversity: numerous small entities scaling up to very few large ones. Old mature forests tend towards this ideal "all size" distribution [47] and are more productive and diverse. It could be that a distribution supporting the greatest diversity is also the most equitable, and makes the most efficient use and mobilization of resources; but this hypothesis needs to be thoroughly examined and tested.

Another key feature (already mentioned) is that activities requiring energy or material are coupled to those that generate them; *and the giving and taking can be reversed as the need arises.* In other words, the system maximizes symbiosis, reciprocity and cooperation. *It is fair and just reciprocal exchange of materials and energy that maintains and ensures the survival of the whole*. Perhaps this is the biggest lesson for the commercial sector that has hitherto operated on exploitation and competition.

This raises the question on the medium of exchange. In human economy, money has become the medium of exchange for goods and services, replacing trust and goodwill. Goods and services have their obvious counterparts in living systems as materials and energy. But what is the equivalent of money in living systems? Is there such a thing?

11. ENERGY IN LIVING SYSTEMS *VERSUS* MONEY IN HUMAN ECONOMY

Energy in living systems is all too often equated with money in economy, and *vice versa*. This has misled generations of biologists and economists alike. As we have seen, living energy transactions are always tied to real interactions and exchanges between molecules with minimum dissipation. For money to be equivalent to energy, it too, has to be tied to exchanges in real goods and services. Unfortunately, money has become more and more decoupled from real goods and services, and hence more like entropy than energy. It becomes entropy in unequal exchanges when the price paid is too low, as in exploitative labor or between currencies that are grossly overvalued on the side of the purchaser. The money market is especially entropic [25,44,48], because it is not based on any real-valued goods or services; furthermore, it artificially inflates the purchasing power of the rich, leading to greater exploitation of environmental resources. This becomes more explicit when the medium for energy exchange in the living system is considered.

The medium of exchange in the living system is adenosine triphosphate (ATP), which many cell biologists and biochemists too, have equated with money. It is the universal energy transducer in living organisms, so much so that its concentration in cells is maintained constant as far as possible by a reserve molecule creatine phosphate, which is allowed to deplete, and its sole function appears to be to keep the supply of ATP constant. ATP has a very special role in maintaining the living state (see [13]).

ATP itself is synthesized using energy from food stuffs or from sunlight, and *never from nothing*. In other words, ATP, the medium of exchange of goods and services in the living system is never decoupled from material and energy cycles, and that is where the analogy to money ends, at least in the current economic system.

In our current human economy, money is being created electronically practically out of nothing, without limit or control; or worse, out of toxic debt and debt pyramids [49]. It has become completely decoupled from real goods and services, and generates real entropy in Earth's ecosystem (and contributing to global warming) when used for overconsumption and exploitation of natural resources. That is ultimately the reason why we need to "shut down Wall Street" [50] in favor of restoring the real high street economy.

Contrary to what many economists assume, an economy that genuinely supports livelihood is not just about the flow of money; it is about people making a living transforming and exchanging materials and energy, and is therefore closely aligned with thermodynamics. Money is the means for the exchange of real goods and services, and not the end of economy. It must never be decoupled from the value of real goods and services. Major disasters like the current financial crisis come about because money has been created without control, and because people mistake the means for the end.

The circular thermodynamics of the living state—the science of energy and material transformation in living organisms and sustainable systems—gives ample support to the following features of a truly green economy [44].

The green economy is a renewable, closed-loop resource use model that includes agriculture, the energy, construction, manufacturing, and service industries, as well as finance. It is a complete way of life

Like nature, the truly green economy maximizes diversity, reciprocity, symbiosis, cooperation and *equity*; greed and inequity are unsustainable

A truly green economy is embedded within nature's economy

It regards nature as the ultimate source of wealth, of "natural capital" [51], which must be regenerated and indeed, increased, in order to feed all sectors of the human economy.

12. CONCLUSIONS

I have presented a circular thermodynamics of organisms and sustainable systems based on dynamic closures in nested space-time domains that enable the system to approach the ideal of zero entropy production simultaneously at equilibrium and far from equilibrium conditions.

A sustainable system, like an organism, maximizes cyclic, non-dissipative flows while minimizing dissipative flows, *i.e.*, it tends towards minimum entropy production. Activities yielding energy are directly coupled to those requiring energy and the direction can be reversed as the need arises. This dynamic space-time differentiation is based on symbiosis, cooperation and reciprocity. It also explains why biodiverse systems are more resilient and productive.

As a model of sustainable systems, it disposes of some major myths in ecology and economy. For example, the carrying capacity of a piece of land is not a constant; it depends on how it is organized for production. Similarly, a sustainable system does not mean one without growth. On the contrary, it can easily grow by engaging more cycles. Finally, it exposes the weaknesses of the dominant economic model of unlimited competitive growth based on money decoupled from real value of goods and services.

ACKNOWLEDGMENTS

Kenneth Denbigh's *Thermodynamics of the Steady State* was a key to theory presented here. I have met and exchanged correspondences with him on the subject during the years after his retirement until shortly before his death. I owe a great deal to the kindness, generosity and grace with which he encouraged my extension of his work. Robert Ulanowicz contacted me out of the blue and coauthored an important paper with me, imparting his considerable wisdom and clarified my thinking enormously; and he too, allowed me graciously to take him where he would not have gone on his own. Environmental engineer George Chan's work on integrated food and waste management was instrumental in the final development of the zero-entropy ideal model of organisms and sustainable systems. My husband and mathematical adviser Peter Saunders is a constant source of support and inspiration, without whom none of this would have been possible.

REFERENCES

1. Ho, M.W. *The Rainbow and the Worm, the Physics of Organisms*, 3rd ed.; World Scientific: London, UK, 2008.

2. Ho, M.W. Are Economic Systems Like Organisms? In *Evolutionary Biology and Economic Theory*; Koslowski, P., Ed.; Springer-Verlag: Berlin, Germany, 1999; pp. 237–258.

3. Ho, M.W.; Ulanowicz, R.E. Sustainable systems as organisms? *BioSystems* **2005**, *82*, 39–51.

4. Ho, M.W. The computer aspires to the human brain. *Sci. Soc.* **2013**, *58*, 14–16.

5. Ho, M.W. Quantum computing or red flag? *Sci. Soc.* **2008**, *42*, 31–33.

6. Ruch, P.; Brunschwiler, T.; Escher, W.; Paredes, S.; Michel, B. Toward five-dimensional scaling: How density improves efficiency in future computers. *IBM J. Res. Dev.* **2011**, *55*.

7. Fischetti, M. Computers *versus* Brains. Available online: http://www.scientificamerican.com/article.cfm?id=computers-vs-ological (accessed on 19 June 2013).

8. IBM Sequoia. Wikipedia. Available online: http://en.wikipedia.org/wiki/IBM_Sequoia (accessed on 10 January 2013).

9. Lobosco, K. China Builds Fastest Supercomputer in the World. Available online: http://money.cnn.com/2013/06/06/technology/enterprise/fastest-supercomputer/ (accessed on 7 June 2013).

10. Ho, M.W.; Lawrence, M. Interference colour vital imaging—A novel non-invasive technique. *Micros. Anal.* **1993**,*September*, 26. [Google Scholar]

11. Ho, M.W.; Haffegee, J.; Newton, R.; Zhou, Y.M.; Bolton, J.S.; Ross, L. Organisms as polyphasic liquid crystals.*Bioelectrochem. Bioenerg.* **1996**, *41*, 81–91.

12. Ross, S.; Newton, R.; Zhou, Y.M.; Haffegee, J.; Ho, M.W.; Bolton, J.P.; Knight, D. Quantitative image analysis of birefringent material. *J. Microsc.* **1997**, *187*, 72–67.

13. Ho, M.W. *Living Rainbow H_2O*; World Scientific & Imperial College Press: London, UK, 2012.

14. Ho, M.W.; Zhou, Y.-M.; Haffegee, J.; Watton, A.; Musumeci, F.; Privitera, G.; Scordino, A. The Liquid Crystalline Organism and Biological Water. In *Water in Cell Biology*; Pollack, G., Ed.; Springer: Dordrecht, Germany, 2006; pp. 219–234.

15. Morowitz, H.J. *Energy Flow in Biology*; Academic Press: New York, NY, USA, 1968.

16. Ho, M.W. *Bioenergetics, Book Two, S327 Living Processes, An Open University Third Level Course*; Open University Press: Milton Keynes, UK, 1995.

17. Schrödinger, E. *What is Life?* Cambridge University Press: Cambridge, UK, 1944.

18. Prigogine, I. *Introduction to Thermodynamics of Irreversible Processes*; John Wiley & Sons: New York, NY, USA, 1967.

19. Ulanowicz, R.E. Identifying the structure of cycling in ecosystems. *Math. Biosci.* **1983**, *65*, 210–237.

20. Onsager, L. Reciprocal relations in irreversible processes. I. *Phys. Rev.* **1931**, *37*, 405–426.

21. Stewart, I. *Does God Play Dice? The New Mathematics of Chaos*; Penguin Books: London, UK, 2002.

22. McClare, C.W.F. Chemical machines, Maxwell's demon and living organisms. *J. Theor. Biol.* **1971**, *30*, 1–34.

23. Ho, M.W. The real bioinformatics revolution, proteins and nucleic acids singing to one another. *Sci. Soc.* **2007**, *33*, 42–45.

24. Ling, G. *Life at the Cell and Below-Cell Level*; Pacific Press: New York, NY, USA, 2001.

25. Ho, M.W. *Genetic Engineering Dream or Nightmare?* Third World Network: New York, NY, USA, 2007.

26. Panchelyuga, V.A.; Panchelyuga, M.S. Resonance and fractals on the real numbers set. *Prog. Phys.* **2012**, *4*, 48–53.

27. Aon, M.; Cortassa, S. Hypothesis on the fractal nature of cytoplasm. *FEBS Lett.* **1994**, *344*, 1–4.

28. Ho, M.W. The heartbeat of health. *Sci. Soc.* **2007**, *35*, 10–13.

29. Ho, M.W. Happiness is a heartbeat away. *Sci. Soc.* **2007**, *35*, 14–18.

30. Odum, E.P. The strategy of ecosystem development. *Science* **1969**, *164*, 262–270.

31. Denbigh, K.G. *The Thermodynamics of the Steady State*; Mathuen Ltd.: New York, NY, USA, 1951.

32. Schneider, E.D.; Kay, J.J. Life as a manifestation of the second law of thermodynamics. *Mathl. Comp. Model.* **1994**, *19*, 25–48.

33. Hannon, B.; Ulanowicz, R.E. Life and the production of entropy. *Proc. R. Soc. Lond. B* **1987**, *232*, 181–192.

34. Annila, A.; Kuismanen, E. Natural hierarchy emerges from energy dispersal. *BioSystems* **2008**, *95*, 227–233.

35. Luvall, J.C.; Holbo, H.R. Thermal Remote Sensing Methods in Landscape Ecology. In *Quantitative Methods in Landscape Ecology*; Turner, M., Gardner, R.H., Eds.; Springer-Verlag: New York, NY, USA, 1991; pp. 127–152.

36. Shiva, V. *Monoculture of the Mind*; Third World Network: Penang, Malaysia, 1993.

37. Schroll, H. Energy-flow and ecological sustainability in Danish agriculture. *Agric. Ecosyst. Environ.* **1994**, *51*, 301–310.

38. Ho, M.W.; Burcher, S.; Lim, L.C. *Food Futures Now, Organic, Sustainable, Fossil Fuel Free*; ISIS/TWN: London, UK and Penang, Malaysia, 2008.

39. Ho, M.W. Sustainable Agriculture and Off-Grid Renewable Energy. In *Wake up before It Is Too Late, Make Agriculture Truly Sustainable Now for Food Security in a Changing Climate*; United Nations Conference on Trade and Development: New York, NY, USA and Geneva, Switzerland, 2013; pp. 72–76.

40. Ho, M.W. Circular economy of the dyke-pond system. *Sci. Soc.* **2006**, *32*, 38–41.

41. Tilman, D.; Reich, P.B.; Knops, J.; Wedin, D.; Mielke, T.; Lehman, C. Diversity and productivity in a long-term grassland experiment. *Science* **2001**, *294*, 843–845.

42. West, B.G.; Brown, J.; Enquist, B. The fourth dimension of life: Fractal geometry and allometric scaling of organisms.*Science* **1999**, *284*, 1677–1679.

43. Ho, M.W. Biology's theory of everything. *Sci. Soc.* **2004**, *21*, 46–47.

44. Ho, M.W. Living, green& circular. *Sci. Soc.* **2012**, *53*, 20–23.

45. Ho, M.W. Sustainable agriculture, green energies & the circular economy. *Sci. Soc.* **2010**, *46*, 8–13.

46. Ho, M.W. Circular economy at Davos. *Sci. Soc.* **2013**, *58*, 10–14.

47. Ho, M.W. Multiple uses of forests. *Sci. Soc.* **2005**, *26*, 18–19.

48. Ho, M.W. Dream Farm 2, Organic, Sustainable and Fossil Fuel Free. In *Food Futures Now, Organic, Sustainable, Fossil Fuel Free*; Ho, M.W., Ed.; ISIS & TWN: London, UK and Penang, Malaysia, 2008; pp. 150–158.

49. Ryan-Collins, J.; Greenham, T.; Werner, R.; Jackson, A. *Where Does Money Come From? A Guide to the UK Monetary and Banking System*; New Economic Foundation: London, UK, 2011.

50. Ho, M.W. Shut down Wall Street! *Sci. Soc.* **2012**, *53*, 16–17, 23.

51. Hawken, P. *The Ecology of Commerce*; Harper Collins: New York, NY, USA, 1993.

10

CHAPTER

Thermodynamics of Binding of Divalent Magnesium and Manganese to Uridine Phosphates: Implications for Diabetes-Related Hypomagnesaemia and Carbohydrate Biocatalysis

Corbin J Zea, Gulden Camci-Unal and Nicola L Pohl*

Department of Chemistry and the Plant Sciences Institute, Iowa State University, Gilman Hall, Ames, IA, 50011-3111, USA

ABSTRACT

Background

Although the necessity of divalent magnesium and manganese for full activity of sugar nucleotidyltransferases and glycosyltransferases is well known, the role of these metal cations in binding the substrates (uridine 5'-triphosphate, glucose-1-phosphate, *N*-acetylglucosamine-1-phosphate, and uridine 5'-diphosphate glucose), products (uridine 5'-diphosphate glucose, uridine 5'-diphosphate *N*-acetylglucosamine, pyrophosphate, and uridine 5'-diphosphate), and/or enzyme is not clearly understood.

Results

Using isothermal titration calorimetry we have studied the binding relationship between the divalent metals, magnesium and manganese, and uridine 5'-phosphates to determine the role these metals play in carbohydrate biosynthesis. It was determined from the isothermal titration calorimetry (ITC) data that Mg^{+2} and Mn^{+2} are most tightly bound to PP_i, K_b = 41,000 ± 2000 M^{-1} and 28,000 ± 50,000 M^{-1} respectively, and UTP, K_b = 14,300 ± 700 M^{-1} and 13,000 ± 2,000 M^{-1} respectively.

Conclusion

Our results indicate that the formal charge state of the phosphate containing substrates determine the binding strength. Divalent metal cations magnesium and manganese showed similar trends in binding to the sugar substrates. Enthalpy of binding values were all determined to be endothermic except for the PP_i case. In addition, entropy of binding values were all found to be positive. From this data, we discuss the role of magnesium and manganese in both sugar nucleotidyltransferase and glycosyltransferase reactions, the differences in metal-bound substrates expected under normal physiological metal concentrations and those of hypomagnesaemia, and the implications for drug design.

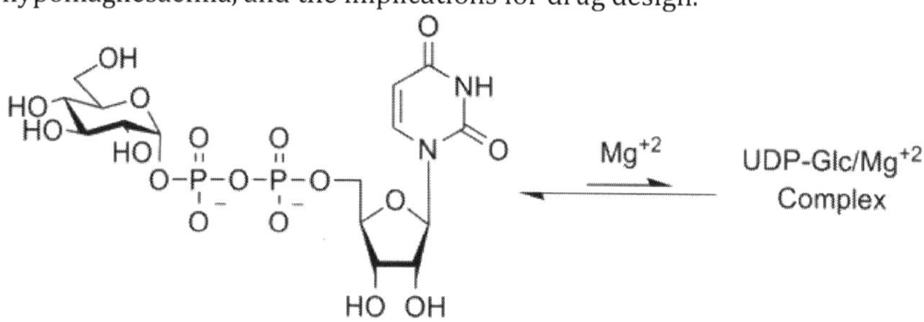

BACKGROUND

The roles that nucleotides play in cellular metabolism range in scope from their use as enzyme substrates to regulators for numerous biochemical pathways. Because of their biological importance and the requirement of divalent metals for biological activity, the binding relationship between various metals and nucleotides has been well documented to shed light on the role of metal-nucleotide complexes in the conformational changes that occur in DNA biosynthesis [1–4] and the energy pathways that involve ATP [5–12].

Recent studies suggest that certain metabolic disorders are linked to serum magnesium concentrations in a way that can be partially explained by these relative metal-binding equilibria. For example, individuals with hypomagnesaemia – reduced serum magnesium concentration – have an increased incidence of metabolic disorders such as type 2 diabetes mellitus [13–20] and high blood pressure [13]. Simple oral administration of magnesium chloride can increase serum magnesium concentrations and decrease total cholesterol and HDL levels while increasing LDL and insulin sensitivity in these patients [21–23]. In patients with high cholesterol and hypomagnesaemia, magnesium treatment is attributed to increasing the effectiveness of statin pharmaceuticals by shifting the equilibrium to the formation of a Mg^{+2}-ATP complex which is a necessary intermediate in the cholesterol biosynthetic pathway [24]. Surprisingly, searches for an explanation of the effect of magnesium supplementation on insulin levels so far have focused only on metal effects at the DNA level [1] and not on the effects of metals on the relative amounts of the substrates involved in glucose metabolism. Such discussions are hindered by the lack of reports of the magnesium binding affinities to these substrates.

The biochemical pathway that synthesizes glycogen and degrades it (Scheme 1) plays an important role in cellular function and metabolism. This pathway is responsible for the storage of glucose and glycogen as well as the regulation of blood glucose levels. For this reason glycogen synthase as well as phosphorylase *b* have been targets of diabetic drug targets [25–28]. It has also been shown that the biocatalyst that activates glucose as ADP-glucose in plants plays a regulatory role in starch biosynthesis [29–31] which tags all three biocatalysts as contributing to the regulation of glucose concentrations.

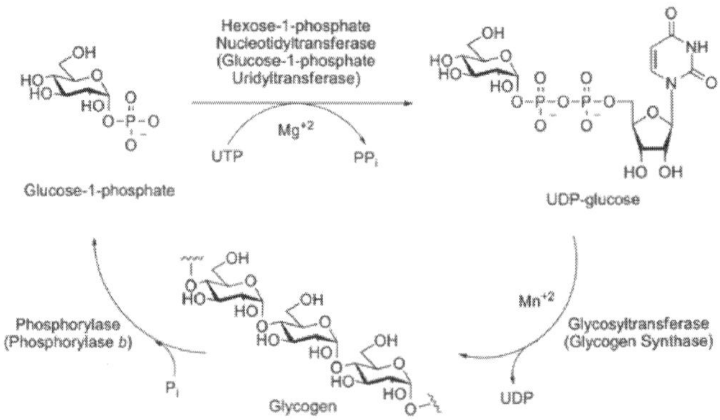

Scheme 1 Biochemical pathway for the synthesis and degradation of glycogen. General classification and the specific names of biocatalysts associated with carbohydrate biosynthesis are shown.

While it is widely known that divalent cations, in particular Mg^{+2} and Mn^{+2}, play a very important role in carbohydrate biocatalysis, the actual mechanism by which the metal interacts with the ligand and the enzymes is not clearly understood. This has important implications in drug development; whether the cation generally binds tightly to the protein active site or binds to the ligand before ligand binding to the protein is an important consideration in designing effective inhibitors of this interaction. It is well documented that uridine triphosphate (UTP), pyrophosphate (PP_i), glucose-1-phosphate (Glc-1-P), and uridine diphosphate glucose (UDP-Glc) all bind and act as inhibitors [32–35], whereas uridine diphosphate (UDP) was demonstrated by Gillett and coworkers [36] to not bind to UDP-Glc pyrophosphorylase. This result is confusing since other researchers have determined UDP to be an inhibitor of UDP-Glc pyrophosphorylase [33, 34]. This raises the question of how inhibitor binding is related to inhibitory affect. Herein, we discuss the binding of Mg^{+2} and Mn^{+2} with UTP, UDP, uridine monophosphate (UMP), UDP-Glc, and PP_i in order to gain an understanding in how and why sugar nucleotidyltransferases and glycosyltransferases (type B) are metal dependant, and to determine in what order metal binding occurs (i.e. does the metal bind to the ligand or biocatalyst first) and what role that has on catalytic activity or inhibition.

RESULTS AND DISCUSSION

While the binding of Mg^{+2} to nucleotide phosphates has been studied previously [5, 7, 11], the binding relationship between the metals, $Mg^{+2}$and Mn^{+2}, and uridine phosphates have not been described. Because the primary interaction between the metal and nucleotides occurs between the phosphates and the metal one would expect the binding parameters between ATP and UTP to be similar. It would also be expected that little binding would be lost by the addition of glucose to UDP to form UDP-glucose. To address these questions, we determined the thermodynamics of binding of Mg^{+2} and Mn^{+2} to UTP, UDP, UMP, UDP-Glc, Glc-1-P, Glc, U, and PP_i at the physiologically relevant temperature of 37°C using isothermal titration calorimetry (ITC) by the direct titration of a solution of magnesium or manganese (II) chloride into a solution containing the phosphate containing compound being studied (Figure 1). It was determined from the ITC data that Mg^{+2} and Mn^{+2} are most tightly bound to PP_i, K_b = 41,000 ± 2000 M^{-1} and 28,000 ± 50,000 M^{-1} respectively, and UTP, K_b = 14,300 ± 700 M^{-1} and 13,000 ± 2,000 M^{-1}respectively (Figure 2). The binding parameters determined for the binding of Mg^{+2} and Mn^{+2} to UDP and UMP were found to be less than that of UTP. The degree of association between Mg^{+2} and Mn^{+2} decreases as phosphates are removed (UTP > UDP > UMP). However, as expected, the binding parameters measured were comparable to previously reported results for the binding of Mg^{+2} to ATP, ADP, AMP, and UMP [1, 5–7]. In addition, as expected, there was no measurable binding of Mg^{+2} and Mn^{+2} to glucose and uridine which do not contain negatively charged phosphates.

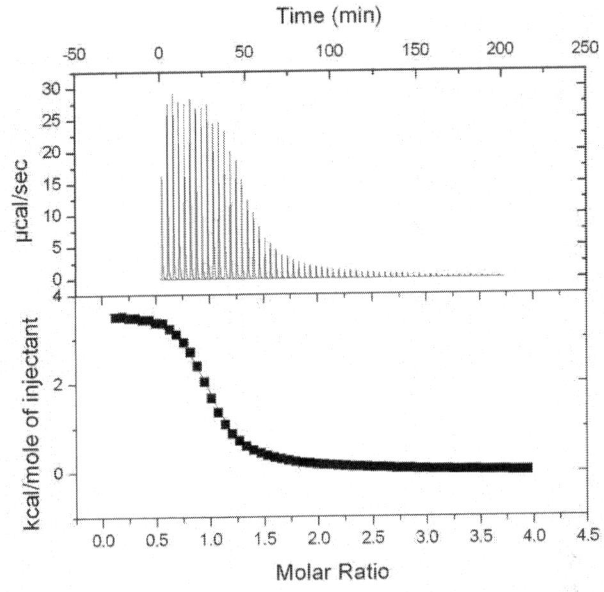

Figure 1 Thermogram (top) and binding isotherm (bottom) showing the addition of 75 mM $MgCl_2$ (syringe) into 2.1 mM UTP (cell) in 100 mM HEPES at pH 7.6 and 37°C.

Comparison of the binding constants for the binding of Mg^{+2} and Mn^{+2} to UDP-Glc to the binding of Mg^{+2} and Mn^{+2} to UDP affords an approximately 10-fold decrease in binding. This significant decrease in binding due to the addition of glucose was not expected; UDP-Glc is often drawn directly bound to these divalent metals. In contrast, the binding values associated with the binding of Mg^{+2} and Mn^{+2} to Glc-1-P were found to be similar in strength to the binding of Mg^{+2} and Mn^{+2} to UMP. This data suggests that the binding of Mg^{+2} and Mn^{+2} to phosphates is dependent on the binding environment of the phosphates involved in the coordination of the metals, Mg^{+2} and Mn^{+2}. For comparison, binding affinities to *N*-

acetylglucosamine, *N*-acetylglucosamine-1-phosphate, and uridine 5'-diphosphate *N*-acetylglucosamine were also measured. From the binding constants the following pattern of affinities emerges: PP_i> UTP > UDP > UMP ≈ UDP-Glc ≈ Glc-1-P ≈ GlcNAc-1-P > UDP-GlcNAc. This data demonstrates a great effect on metal affinities upon substitution at phosphate. If the phosphate has no substituents covalently bound, as in the case of PP_i, the binding constant is large. However, in the case of UMP, UDP-Glc, and Glc-1-P, in which the substrate is bound directly to the binding phosphate, the binding constant is significantly smaller. Finally, in the examples of substrates not having phosphate groups present, GlcNAc, Glc and U, the binding of Mg^{+2} and Mn^{+2} to the substrates was so weak as to preclude accurate binding affinity determinations, which indicates the importance that the charged phosphates play in binding divalent metals.

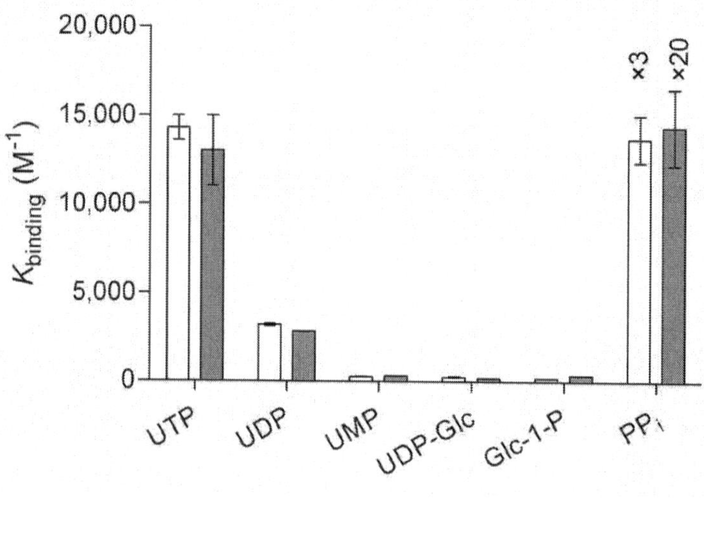

Figure 2. Binding constants of phosphates bound to Mg^{+2} (white) and Mn^{+2} (gray).

Thermodynamic values for the titrations were determined and for all cases, except for the titration of pyrophosphate by Mg^{+2} and Mn^{+2}, the binding enthalpies were endothermic. All titrations afforded a positive change in entropy, which correlates to an increase in disorder caused by the loss of a water shell around the metal and phosphate compound as the metal coordinates to the phosphate compound. This increase in disorder allowed the overall coordination of Mg^{+2} and Mn^{+2} to the phosphate to be thermodynamically allowed with the free energy following a similar pattern as the binding values (PP_i > UTP > UDP > UMP ≈ UDP-Glc ≈ Glc-1-P ≈ GlcNAc-1-P > UDP-GlcNAc).

Although Mg^{+2} and Mn^{+2} bind at very similar strengths to phosphates, the intracellular concentrations of Mg^{+2} and Mn^{+2} vary greatly. Although metal concentrations can vary depending on cell state, type, and organelle, given intracellular concentrations of Mg^{+2} and Mn^{+2} of 1.5 nM [37] and 10 nM [38] respectively, the amount of phosphate expected to be bound to Mg^{+2} and Mn^{+2} varies greatly as shown in Figure 3. At these physiological metal concentrations, the concentration of Mn^{+2} is not large enough to shift the equilibrium of binding in the direction of metal-phosphate complex formation. However, all of the phosphates show a measurable degree of binding to Mg^{+2} at intracellular concentrations of Mg^{+2}. Uridine triphosphate and PP_i are almost completely bound (> 95%) to Mg^{+2} at intracellular concentrations of Mg^{+2}, while UDP is mostly bound (83%), and UMP, UDP-Glc, and Glc-1-P are only partially bound to Mg^{+2} (~ 25%).

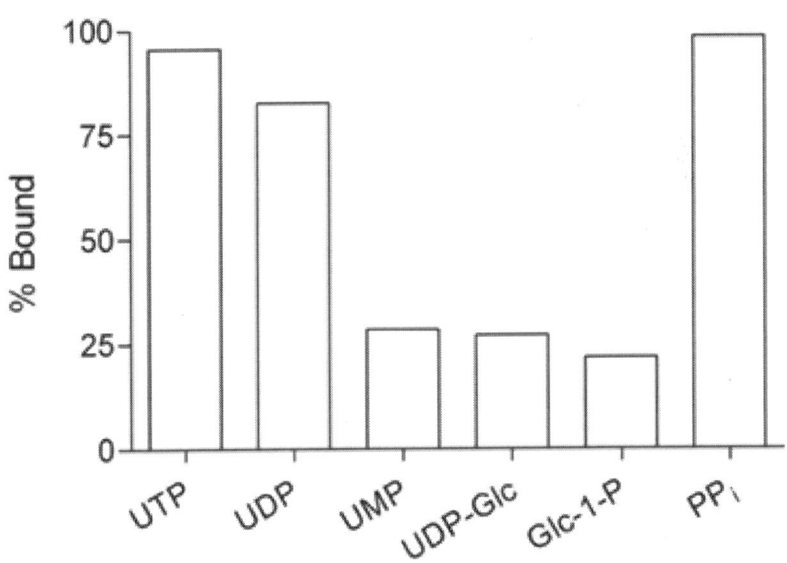

Figure 3. Percent of phosphate bound to Mg^{+2} (white) at the intracellular concentration of 1.5 mM. No measurable binding is observed for Mn^{+2}.

When the concentration of magnesium decreases, as is the case in hypomagnesaemia (0.7 mM Mg^{+2}) there can be significant changes in the phosphate-magnesium concentration (Figure 4). In cases where Mg^{+2} is tightly bound, PP_i and UTP, there is not a large change in complex formation. However, when the binding is not as tight, as is the case with UDP, UMP, UDP-Glc, Glc-1-P, GlcNAc-1-P, and UDP-GlcNAc, the concentration of the phosphate-magnesium complex is greatly reduced by the change in magnesium ion concentration associated with hypomagnesaemia. This decrease in magnesium-phosphate complex formation would be expected to lead to decreased activity of glycogen synthase because of the role that AMP plays as an allosteric activator as well as decreasing the activity of phosphorylase *b* which is allosterically activated by glucose-6-phosphate.

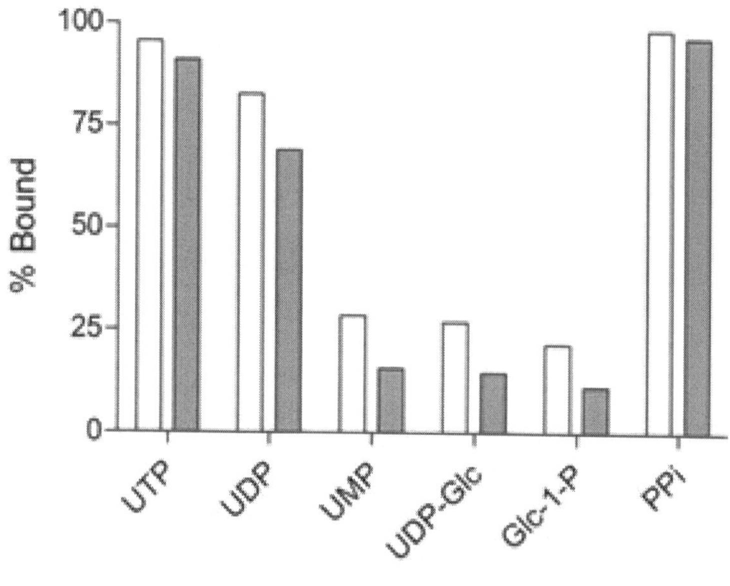

Figure 4. Percent of phosphate bound to Mg^{+2} at a normal physiological Mg^{+2} concentration of 1.5 mM (white) and at a hypomagnesaemia concentration of 0.7 mM (gray).

To understand the role that metal binding plays in enzymatic activity we analyzed several known crystal structures of sugar NTs [39–49], GT-As [50–60] and GT-Bs [61–67]. The active site analysis of sugar NT crystal structures revealed the metal ion, Mg^{+2}, ligands to be one to two oxygen atoms from the phosphates and two to three coordinated oxygens from aspartic acid residues from the enzyme in question [40, 48]. Though not as complex, the coordination of Mg^{+2} in sugar NTs is similar to the octahedral coordination site of Mn^{+2} seen in GT-As, where the active site analysis shows two of the coordinating atoms to Mn^{+2} are oxygens from the α- and β- phosphates and three to four coordination sites are direct interactions to the enzyme [50, 52–54], [56–60]. A well-defined metal coordination site in GT-As allows for a high degree of coordination to occur between the metal, Mn^{+2}, and the protein; this fact implies the presence of a permanent metal binding site [54]. This analysis coupled with the fact of a low intracellular concentration of Mn^{+2} (10 nM) [38] makes the presence of a permanent metal binding site appropriate unless higher local metal concentrations can be maintained at the site of action of particular GTs. The enzymes that constitute GT-Bs, which are not metal dependent, have positively charged amino acid residues, four arginines, which are able to form hydrogen bonds directly to the phosphates [61–67].

CONCLUSION

Our findings of no manganese-phosphate complex formation at physiological manganese ion concentration fits previously reported crystal structure determinations demonstrating the presence of a permanent binding site in GT-As for manganese [54]. For highest activity at low manganese concentration, the enzyme would need to have a way to permanently sequester manganese for full biocatalytic function. However for sugar NTs, which are magnesium dependent, the need for a permanent binding site is not required due to the much higher physiological concentration of divalent magnesium in healthy cells. As for inhibitors of sugar NTs, free UTP has been shown to be an inhibitor of Mg-UTP [35], and explains why UDP is an inhibitor in the presence of Mg^{+2} while it was found not to bind to the enzyme in the absence of a divalent metal [35].

The role magnesium plays in the formation of metal-phosphate complexes may shed some light in why magnesium chloride therapy has lead to a decrease in serum glucose levels in patients with type 2 diabetes [20–23]. With the increase of the serum Mg^{+2} concentrations, there will be an increase in the formation of magnesium-phosphate complexes. The complexes formed play a critical role by activating the biocatalysts involved in the manner in which glucose is biochemically stored and released. The decrease in biocatalyst function would not allow for the biochemical pathway to function properly, adequately sequestering glucose as glycogen when glucose levels are high and degrading glycogen when glucose levels are low. Obviously, these implications await human trials now that the basic biochemical data is available.

Knowledge of the method in which the metal binds to the substrate can also play an important role in drug design. When contemplating inhibitor design for sugar NTs, one must design the inhibitor such that it is able to compete against the most prevalent form of the substrate, which for this class of enzymes is the NTP-Mg^{+2} complex. According to our results, the metal cation binds to NTP first and then after the reaction it leaves bound to PP_i. In contrast, in designing inhibitors for glycosyltransferases, one needs to account for the binding that occurs between the manganese and the substrate because manganese is likely to be permanently bound in the active site and important in the catalytic activity of GT-As while not being important for the catalytic activity of GT-Bs. This knowledge of metal affinities therefore may aid in the development of drug candidates that are able to selectively inhibit one class of glycosyltransferases and phosphorylases over another and ultimately also may lead to diabetic drug development [25–28].

EXPERIMENTAL

2-[4-(2-hydroxyethyl)-1-piperazinyl]-ethanesulfonic acid (Hepes), tetrasodium pyrophosphate decahydrate, magnesium chloride, and manganese (II) chloride were all purchased from Fischer Scientific Company (Hanover Park, IL). All other compounds were purchased from Sigma Chemical Co. (St. Louis, MO) unless otherwise noted and used without further purification. Nanopure water (18.1 MHz) prepared from a Barnstead E-pure water purification system was employed throughout.

Isothermal titration calorimetry

ITC experiments were performed on a Microcal VP-ITC microcalorimeter (Northampton, MA) which was calibrated using the built-in electrical calibration check. All ITC experiments were conducted in 100 mM Hepes buffer pH 7.5 at 37°C. Nucleotide phosphate concentration was determined spectrophotometrically from UV absorbance measurements at 262 nm using an extinction coefficient of ε_{262} = 20,000 cm^{-1} M^{-1}[68]. Glucose-1-phosphate concentrations were determined via electrospray ionization mass spectrometry (ESI-MS) [69]. All solutions were degassed immediately prior to use. Titrant solution (2.5, 7.5, or 10 μL) was added from a 300 μL microsyringe at an interval of 200 sec. into the stirred sample cell (1.4288 mL) containing the nucleotide phosphate at a stirring rate of 480 rpm. To correct for heats of dilution, control experiments were performed by making identical injections of the titrant solution into a cell containing only buffer and these values were subtracted from the titration of the titrant solution into the reaction cell. The titrant solution contained the metal, Mg^{+2} or Mn^{+2} (75 mM), in Hepes buffer (100 mM, pH 7.5). The reaction cell contained 2.5 – 7.5 mM substrate (UTP, UDP, UMP, UDP-Glc, Glc-1-P, Glc, U, or PP_i) in Hepes buffer (100 mM, pH 7.5). Data was analyzed using nonlinear least-squares curve fitting in Origin (7.0, OriginLab Corp., Northampton, MA) using the standard one binding site model supplied by Origin. This analysis yielded the thermodynamic parameters *K* (binding constant), enthalpy of binding (ΔH), entropy of binding (ΔS), and *n*, where *n* is the ratio of the metal ion to the substrate in the complex. See Additional file1 for Isothermal titration calorimetry plots and tables containing *n*, *K*, ΔG, ΔH, and ΔS metal binding information for all substrates and metals and their thermograms.

ABBREVIATIONS

ADP: Adenosine diphosphate

AMP: Adenosine monophosphate

ATP: Adenosine 5'-triphosphate

DNA: Deoxyribonucleic acid

DPA: Diphenylamine reagent

ESI-MS: Electrospray ionization mass spectrometry

G: Free energy of binding [cal/mol]

Glc: Glucose

Glc-1-P: Glucose-1-phosphate

GlcNAc: *N*-Acetylglucosamine

GlcNAc-1-P: *N*-Acetylglucosamine-1-phosphate

GT-A: Glycosyltransferase type A

GTB: Glycosyltransferase type B

H: Enthalpy of binding [cal/mol]

HDL: High-density lipoproteins

ITC: Isothermal titration calorimetry

K: Binding constant [M^{-1}]

LDL: Low-density lipoproteins

n: Ratio of metal to substrate

NT: Nucleotidyltransferase

S: Entropy of binding [cal/mol·T]

U: Uridine

UDP: Uridine diphosphate

UDP-Glc: UDP-glucose

UDP-GlcNAc: Uridine diphosphate-*N*-Acetylglucosamine

UMP: Uridine monophosphate

UTP: Uridine triphosphate

UV: Ultraviolet spectroscopy

ACKNOWLEDGEMENTS

We thank V. Frasca (Microcal) for helpful discussions and the Department of Chemistry, the Plant Sciences Institute, and the Herman Frasch Foundation (American Chemical Society) for partial support of this research. This material is based in part upon work supported by the National Science Foundation under CAREER Grant No. 0349139. N.L.P. is a Cottrell Scholar of Research Corporation and an Alfred P. Sloan Research Fellow.

AUTHORS' CONTRIBUTIONS

CJZ and GC-U collected ITC data, performed data analysis, and contributed to drafting the manuscript. NLP conceived of the study, participated in its design and coordination and contributed to drafting the manuscript. All authors read and approved the final manuscript.

REFERENCES

1. Herrero LA, Terron A: Complexation in solution of magnesium (II) and cobalt (II) with purine 5'-monophosphates and pyrimide 5'-monophosphates: a potentiometric and calorimetric study. Polyhedron. 1998, 17: 3825-3833.
2. Hartwig A: Role of magnesium in genomic stability. Mutat Res. 2001, 475: 113-121.
3. Barton JK: Metal/Nucleic Acid Interactions. Bioinorganic Chemistry. Edited by: Bertini I, Gray HB, Lippard SJ. 1994, Valentine JS: University Science Books, 505-515.
4. Sigel H: Metal ion complexes of antivirally active nucleotide analogues. Conclusions regarding their biological action. Chem Soc Rev. 2004, 33: 191-200. 10.1039/b310349h.
5. Shanbhag SM, Choppin GR: Thermodynamics of Mg and Ca complexation with AMP, ADP, ATP. Inorg Chim Acta. 1987, 138: 187-192.
6. Massoud SS, Sigel H: Metal Ion Coordinating Properties of Pyrimidine-Nucleoside 5'-Monophosphates (CMP, UMP, TMP) and of Simple Phosphate Monoesters, Including D-Ribose 5'-Monophosphate. Establishment of Relations between Complex Stability and Phosphate Basicity. Inorg Chem. 1988, 27: 1447-1453.
7. Wilson JE, Chin A: Chelation of Divalent Cations by ATP, Studies by Titration Calorimetry. Anal Biochem. 1991, 193: 16-19..
8. Sigel H: Metal ion-assisted stacking interactions and the facilitated hydrolysis of nucleoside 5'-triphosphates. Pure Appl Chem. 1998, 70: 969-976.
9. Herrero LA, Terron A: Interactions in solution of calcium (II) and copper (II) with nucleoside monophosphates: a calorimetric study. J Biol Inorg Chem. 2000, 5: 269-275. 10.1007/s007750050371.
10. Stumber M, Herrmann C, Wohlgemuth S, Kalbitzer HR, Jahn W, Geyer M: Synthesis, characterization and application of two nucleoside triphosphate analogues, GTPcNH and GTPcF. Eur J Biochem. 2002, 269: 3270-3278.
11. Bianchi EM, Sajadi SAA, Song B, Sigel H: Stabilities and Isomeric Equilibria in Aqueous Solution of Monomeric Metal Ion Complexes of Adenosine 5'-Diphosphate (ADP^{3-}) in Comparison with Those of

Adenosine 5'-Monophosphate (AMP^{2-}). Chem Eur J. 2003, 9: 881-892.

12. Vogel HJ, Bridger WA: Phosphorus-31 Nuclear Magnetic Resonance Studies of the Methylene and Fluoro Analogues of Adenine Nucleotides. Effects of pH and Magnesium Ion Binding. Biochemistry. 1982, 21: 394-401.

13. Guerrero-Romero F, Rodriguez-Moran M: Low serum magnesium levels and metabolic syndrome. Acta Diabetol. 2002, 39: 209-213.

14. Volpe SM: Magnesium, the Metabolic Syndrome, Insulin Resistance, and Type 2 Diabetes Mellitus. Crit Rev Food Sci Nutr. 2008, 48: 293-300.

15. Wolf FI, Trapani V: Cell (patho)physiology of magnesium. Clinical Science. 2008, 114: 27-35.

16. Pham P-CT, Pham P-MT, Pham SV, Miller JM, Pham P-TT: Hypomagnesemia in Patients with Type 2 Diabetes. Clin J Am Soc Nephrol. 2007, 2: 366-373.

17. Barbagallo M, Dominguez LJ: Magnesium metabolism in type 2 diabetes mellitus, metabolic syndrome and insulin resistance. Arc Biochem Biophys. 2007, 458: 40-47..

18. Sales CH, Pedrosa LFC: Magnesium and diabetes mellitus: Their relation. Clin Nutr. 2006, 25: 554-562..

19. Schroder H: Protective mechanisms of the Mediterranean diet in obesity and type 2 diabetes. J Nutr Biochem. 2007, 18: 149-160.

20. Guerrero-Romero F, Rodriguez-Moran M: Hypomagnesemia, oxidative stress, inflammation, and metabolic syndrome. Diabetes Metab Res Rev. 2006, 22: 471-476. 10.1002/dmrr.644.

21. Rodriguez-Moran M, Guerrero-Romero F: Oral magnesium supplementation improves insulin sensitivity and metabolic control in type 2 diabetic subjects: a randomized double-blind controlled trial. Diabetes Care. 2003, 26: 1147-1152.

22. Guerrero-Romero F, Tamez-Perez HE, Gonzalez-Gonzalez G, Salinas-Martinez AM, Montes-Villarreal J, Trevino-Ortiz JH, Rodriguez-Moran M: Oral Magnesium supplementation improves insulin sensitivity in non-diabetic subjects with insulin resistance. A double-blind placebo-controlled randomized trial. Diabetes Metab. 2004, 30: 253-258.

23. Lopez-Ridaura R, Willett WC, Rimm EB, Liu S, Stampfer MJ, Manson JE, Hu FB: Magnesium intake and risk of type 2 diabetes in men and women. Diabetes Care. 2004, 27: 134-140.

24. Rosanoff A, Seelig MS: Comparison of mechanism and functional effects of magnesium and statin pharmaceuticals. J Am Coll Nutr. 2004, 23: 501S-505S.

25. Martin WH, Hoover DJ, Armento SJ, Stock IA, McPherson RK, Danley DE, Stevenson RW, Barrett EJ, Treadway JL: Discovery of a human liver glycogen phosphorylase inhibitor that lowers blood glucose in vivo. Proc Natl Acad Sci USA. 1998, 95: 1776-1781..

26. Oikonomakos NG, Schnier JB, Zographos SE, Skamnaki VT, Tsitsanou KE, Johnson LN: Flavopiridol inhibits glycogen phosphorylase by binding at the inhibitor site. J Biol Chem. 2000, 275: 34566-34573. 10.1074/jbc.M004485200.

27. Treadway JL, Mendys P, Hoover DJ: Glycogen phosphorylase inhibitors for treatment of type 2 diabetes mellitus. Expert Opin Invest Drugs. 2001, 10: 439-454..

28. Oikonomakos NG, Zographos SE, Skamnaki VT, Archontis G: The 1.76 A resolution crystal structure of glycogen phosphorylase b complexed with glucose, and CP32 a potential antidiabetic drug. Bioorg Med Chem. 0626, 10: 1313-1319.

29. Green TA, Hannah LC: Adenosine diphosphate glucose pyrophosphorylase, a rate-limiting step in starch biosynthesis. Physiologia Plantarum. 1998, 103: 574-580.

30. Ballicora MA, Iglesias AA, Preiss J: ADP-glucose pyrophosphorylase, a regulatory enzyme for bacterial glycogen synthesis. Microbiol Mol Biol Rev. 2003, 67: 213-225.

31. Ballicora MA, Iglesias AA, Preiss J: ADP-Glucose Pyrophosphorylase: A Regulatory Enzyme for Plant Starch Synthesis. Photosynth Res. 2004, 79: 1-24.

32. Bernstein RL, Robbins PW: Control Aspects of Uridine 5'-Diphosphate Glucose and Thymidine 5'-Diphosphate Glucose Synthesis by Microbial Enzymes. J Biol Chem. 1965, 240: 391-397.

33. Albrecht GJ, Bass ST, Seifert LL, Hansen RG: Crystallization and Properties of Uridine Diphosphate Glucose Pyrophosphorylase from Liver. J Biol Chem. 1966, 241: 2968-2975.

34. Roach PJ, Warren KR, Atkinson DE: Uridine Diphosphate Glucose Synthase from Calf Liver: Determinants of Enzyme Activity in Vitro. Biochemistry. 1975, 14: 5445-5450.

35. Elling L: Kinetic Characterization of UDP-Glucose Pyrophosphorylase from Germinated Barley (Malt). Phytochem. 1996, 42: 955-960.

36. Gillett TA, Levine S, Hansen RG: Uridine Diphosphate Glucose Pyrophosphorylase. J Biol Chem. 1971, 246: 2551-2554.

37. Csernoch L, Bernengo JC, Szentesi P, Jacquemond V: Measurements of Intracellular MgConcentration in Mouse Skeletal Muscle Fibers with the Fluorescent Indicator Mag-Indo-1. Biophys J. 1998, 75: 957-967.

38. Cowan JA: Inorganic Biochemistry: An Introduction. 1997, VCH Publishers Inc

39. Brown K, Pompeo F, Dixon S, Mengin-Lecreulx D, Cambillau C, Bourne Y: Crystal structure of the bifunctional *N*-acetylglucosamine 1-phosphate uridyltransferase from *Escherichia coli*: a paradigm for the related pyrophosphorylase superfamily. EMBO J. 1999, 18: 4096-4107.

40. Blankenfeldt W, Asuncion M, Lam JS, Naismith JJ: The structural basis of the catalytic mechanism and regulation of glucose-1-phosphate thymydylyltransferase (RmlA). EMBO J. 2000, 19: 6652-6663.

41. Barton WA, Lesniak J, Biggins JB, Jeffrey PD, Jiang J, Rajashankar KR, Thorson JS, Nikolov DB: Structure, mechanism and engineering of a nucleotidylyltransferase as a first step toward glycorandomization. Nature Struct Biol. 2001, 8: 545-551. 10.1038/88618.

42. Kostrewa D, D'Arcy A, Takacs B, Kamber M: Crystal Structures of Streptococcus pneumoniae N-Acetylglucosamine-1-phosphate Uridyltransferase, GlmU, in Apo Form at 2.33A Resolution and in Complex with UDP-N-Acetylglucosamine and Mg^{2+} at 1.6A Resolution. J Mol Biol. 2001, 305: 279-289. 10.

43. Olsen LR, Roderick SL: Structure of the Escherichia coli GlmU Pyrophosphorylase and Acetyltransferase Active Sites. Biochemistry. 2001, 40: 1913-1921. 10.

44. Peneff C, Ferrari P, Charrier V, Taburet Y, Monnier C, Zamboni V, Winter J, Harnois M, Fassy F, Bourne Y: Crystal structures of two

human pyrophosphorylase isoforms in complexes with UDPGlc(Gal)NAc: role of the alternatively spliced insert in the enzyme oligomeric assembly and active site architecture. EMBO J. 2001, 20: 6191-6202. 10.

45. Sayre PH, Finer-Moore JS, Fritz TA, Biermann D, Gates SB, MacKellar WC, Patel VF, Stroud RM: Multi-targeted Antifolates Aimed at Avoiding Drug Resistance Form Covalent Closed Inhibitory Complexes with Human and Escherichia coli Thymidylate Synthases. J Mol Biol. 2001, 313: 813-829. 10.

46. Sulzenbacher G, Gal L, Peneff C, Fassy F, Bourne Y: Crystal Structure of Streptococcus pneumoniae N-Acetylglucosamine-1-phosphate Uridyltransferase Bound to Acetyl-coenzyme A Reveals a Novel Active Site Architecture. J Biol Chem. 2001, 276: 11844-11851. 10.

47. Zuccotti S, Zanardi D, Rosano C, Sturla L, Tonetti M, Bolognesi M: Kinetic and Crystallographic Analyses Support a Sequential-ordered Bi Bi Catalytic Mechanism for Escherichia coli Glucose-1-phosphate Thymidylyltransferase. J Mol Biol. 2001, 313: 831-843. 10..

48. Sivaraman J, Sauve V, Matte A, Cygler M: Crystal Structure of *Escherichia coli* Glucose-1-Phosphate Thymidylyltransferase (RffH) Complexed with dTTP and Mg^{2+}. J Biol Chem. 2002, 277: 44214-44219. 10.

49. Thoden JB, Ruzicka FJ, Frey PA, Rayment I, Holden HM: Structural Analysis of the H166G Site-Directed Mutant of Galactose-1-phosphate Uridylyltransferase Complexed with either UDP-glucose or UDP-galactose: Detailed Description of the Nucleotide Sugar Binding Site. Biochemistry. 1997, 36: 1212-1222. 10..

50. Charnock SJ, Davies GJ: Structure of the Nucleotide-Diphospho-Sugar Transferase, SpsA from *Bacillus subtilis*, in Native and Nucleotide-Complexed Forms. Biochemistry. 1999, 38: 6380-6385. 10.1021/bi990270y.

51. Gastinel LN, Cambillau C, Bourne Y: Crystal structures of the bovine b4galactosyltransferase catalytic domain and its complex with uridine diphosphogalactose. EMBO J. 1999, 18: 3546-3557. 10..

52. Unligil UM, Zhou S, Yuwaraj S, Sarkar M, Schachter H, Rini JM: X-Ray Crystal Structure of Rabbit N-Acetylglucosaminyltransferase I: catalytic mechanism and a new protein superfamily. EMBO J. 2000, 19: 5269-5280. 10.

53. Gastinel LN, Bignon C, Misra AK, Hindsgaul O, Shaper JH, Joziasse DH: Bovine α1,3-galactosyltransferase catalytic domain structure and its relationship with ABO histo-blood group and glycosphingolipid glycosyltransferases. EMBO J. 2001, 20: 638-649. 10.

54. Persson K, Ly HD, Dieckelmann M, Wakarchuk WW, Withers SG, Strynadka NCJ: Crystal structure of the retaining galactosyltransferase LgtC from *Neisseria meningitidis* in complex with donor and acceptor sugar analogs. Nature Struct Biol. 2001, 8: 166-175. 10.

55. Boix E, Zhang Y, Swaminathan GJ, Brew K, Acharya KR: Structural Basis of Ordered Binding of Donor and Acceptor Substrates to the Retaining Glycosyltransferase, α-1,3-Galactosyltransferase. J Biol Chem. 2002, 277: 28310-28318. 10.1074/jbc.M202631200.

56. Gibbons BJ, Roach PJ, Hurley TD: Crystal Structure of the Autocatalytic Initiator of Glycogen Biosynthesis, Glycogenin. J Mol Biol. 2002, 319: 463-477. 10.

57. Patenaude SI, Seto NO, Borisova SN, Szpacenko A, Marcus SL, Palcic MM, Evans SV: The structural basis for specificity in human ABO(H) blood group biosynthesis. Nature Struct Biol. 2002, 9: 685-690. 10.1038/nsb832

58. Pedersen LC, Darden TA, Negishi M: Crystal Structure of β1,3-Glucuronyltransferase I in Complex with Active Donor Substrate UDP-GlcUA. J Biol Chem. 2002, 277: 21869-21873. 10.

59. Ramakrishnan B, Balaji PV, Qasba PK: Crystal Structure of β1,4-Galactosyltransferase Complex with UDP-Gal Reveals an Oligosaccharide Acceptor Binding Site. J Mol Biol. 2002, 318: 491-502. 10.

60. Negishi M, Dong J, Darden TA, Pedersen LG, Pedersen LC: Glucosaminylglycan biosynthesis: what we can learn from the X-ray crystal structures of glycosyltransferases GlcAT1 and EXTL2. Biochem Biophys Res Commun. 2003, 303: 393-398. 10..

61. Vrielink A, Ruger W, Driessen HPC, Freemont PS: Crystal structure of the DNA modifying enzyme β-glucosyltransferase in the presence and absence of the substrate uridine diphosphoglucose. EMBO J. 1994, 13: 3413-3422.

62. Morera S, Imberty A, Aschke-Sonnenborn U, Ruger W, Freemont PS: T4 Phage β-Glucosyltransferase: Substrate Binding and Proposed Catalytic Mechanism. J Mol Biol. 1999, 292: 717-730.

63. Ha S, Walker D, Shi Y, Walker S: The 1.9 Å crystal structure of Escherichia coli MurG, a membrane-associated glycosyltransferase involved in peptidoglycan biosynthesis. Protein Sci. 2000, 9: 1045-1052.

64. Morera S, Lariviere L, Kurzeck J, Aschke-Sonnenborn U, Freemont PS, Janin J, Ruger W: High Resolution Crystal Structures of T4 Phage β-Glucosyltransferase: Induced Fit and Effect of Substrate and Metal Binding. J Mol Biol. 2001, 311: 569-577. 10.1006/jmbi.2001.4905.

65. Hu Y, Chen L, Ha S, Gross B, Falcone B, Walker D, Mokhtarzadeh M, Walker S: Crystal structure of the MurG:UDP-GlcNAc complex reveals common structural principles of a superfamily of glycosyltransferases. Proc Natl Acad Sci USA. 2003, 100: 845-849.

66. Lariviere L, Gueguen-Chaignon V, Morera S: Crystal Structures of the T4 Phage β-Glucosyltransferase and the D100A Mutant in Complex with UDP-glucose: Glucose Binding and Identification of the Catalytic Base for a Direct Displacement Mechanism. J Mol Biol. 2003, 330: 1077-1086. 10.

67. Mulichak AM, Lu W, Losey HC, Walsh CT, Garavito RM: Crystal Structure of Vancosaminyltransferase GtfD from the Vancomycin Biosynthetic Pathway: Interactions with Acceptor and Nucleotide Ligands. Biochemistry. 2004, 43: 5170-5180.

68. O'Neil MJ, Smith A, Heckelman PE, Obenchain JR, Gallipeau JAR, D'Arecca MA, Budavari S: The Merck Index. Merck. 2001

69. Zea CJ, Pohl NL: Kinetic and substrate binding analysis of phosphorylase b via electrospray ionization mass spectrometry: a model for chemical proteomics of sugar phosphorylases. Anal Biochem. 2004, 327: 107-113.

11
CHAPTER

First-Principles Calculations of the Structural, Mechanical and Thermodynamics Properties of Cubic Zirconia

Ibrahim D. Muhammad[1*], Mokhtar Awang[1], Othman Mamat[1], Zilati Bt Shaari[2]

[1]*Mechanical Engineering Department, Universiti Teknologi Petronas, Seri Iskandar, Malaysia*

[2]*Chemical Engineering Department, Universiti Teknologi Petronas, Seri Iskandar, Malaysia*

ABSTRACT

The structural, mechanical and thermodynamics properties of cubic zirconium oxide ($cZrO_2$) were investigated in this study using ab initio or first-principles calculations. Density functional theory was used to optimize the crystal structure of $cZrO_2$ and thereafter, simulations were conducted to predict the lattice parameters and elastic constants. The Zr-O bond distance was calculated as 2.1763 Å with unit cell density of 6.4179 g/cm^3. The data obtained were used to determine Young's modulus, bulk modulus, Poisson's ratio and hardness of $cZrO_2$ as 545.12 GPa, 136.464 GPa, 0.1898 and 12.663(H_v) respectively. The result indicates that $cZrO_2$ is mechanically stable with thermodynamics properties of a refractory material having potential for structural and catalytic applications in various forms as a nanomaterial.

Keywords: Cubic Zirconium Oxide, First-Principles Calculation, CASTEP, Elastic Constants

1. INTRODUCTION

Recently, zirconium oxide in cubic polymorph has attracted interest due to its mechanical and thermal properties. These interests have resulted in the use of $cZrO_2$ to various applications such as catalysts, fuel cells, oxygen sensors and others [1] -[4] . Thus $cZrO_2$ has been synthesized at nanoscale in various forms. The sol-gel method was used to produce ZrO_2 nanoparticles with fairly uniform dimension ranging from 50 nm to 90 nm [5] . ZrO_2 nanosheets with thickness in the range of 3.2 - 4.2 nm were produced through bottom-up synthesis by impregnation of graphene oxide in cyclohexane containing Zr-based alkoxides [6] . Also, ZrO_2 nanotubes have been prepared via several methods such as direct anodization, template-assisted depositions and hydrothermal treatments [2] ; the nanotubes obtained have different inner and outer diameters, thicknesses and lengths.

For efficient and cost-effective applications of $cZrO_2$, its structural and mechanical properties are required at molecular and atomic scale. But due to complexity and high cost of equipment, only limited experiments have been conducted to characterize required properties of $cZrO_2$ at nanoscale and/or atomic level [2] [4] . This led to computational modeling and simulation of various properties of $cZrO_2$ based on atomistic and continuum approaches [1] [2] . However, most of the simulations already conducted are limited to properties such as electronic structures, phonon dispersion, surface adsorption and diffusion [7] -[13] with limited emphasis on structural and mechanical properties.

Therefore, in this study the first-principles calculations are utilized to numerically predict the structural characteristics, mechanical properties and thermodynamic properties of $cZrO_2$. The values obtained are used to analyze the structural stability of $cZrO_2$ and compared available data for further simulation(s).

2. COMPUTATIONAL METHODS

The 3D structure of $cZrO_2$ was modeled from available data [2] using the surface builder tools of the Material Studio software [14] . The crystal structure of$cZrO_2$ was optimized geometrically in order to obtain initial lattice parameters and density based on density function theory (DFT) as implemented in Cambridge Sequential Total Energy Package (CASTEP) code [15] . In this calculation, the Generalized Gradient Approximation (GGA) having Burke-ErnZerhof Potential (PBE) for solids was used [13] . To obtain accurate structures, the calculations were conducted in the irreducible Brillouin zone with 8 × 8 × 8 k point mesh Monk horst-Pack scheme [16] . In order to obtain plane wave expansions, a kinetic energy cut-off value of 380 eV was used. Thereafter the BroydenFletcher-Goldfarb-Shannon (BFGS) optimization method was used with fixed basis quality to obtain the symmetric crystal structure of $cZrO_2$. During the optimization process, the total energy was designed to converge to 5×10^{-6} eV and the force per atom diminished to 0.002 eV/Å. The optimized structure of $cZrO_2$ was then simulated to obtain required mechanical and thermodynamics properties.

3. RESULTS AND DISCUSSION

3.1. Crystal Structure

The bulk crystal structure of $cZrO_2$ belongs to the Fm-3m space [4] [9] and is modeled in various forms as seen in Figure 1. The geometric structure has influence on the properties of the material even at nanoscale [2] . After optimization, all the lattice parameters were recorded and compared with values obtained from other simulations and experiments as summarized in Table1 It was observed that the deviation of the lattice parameters is less than 3% from the experimental values, which may be due to the approximation method used during optimization.

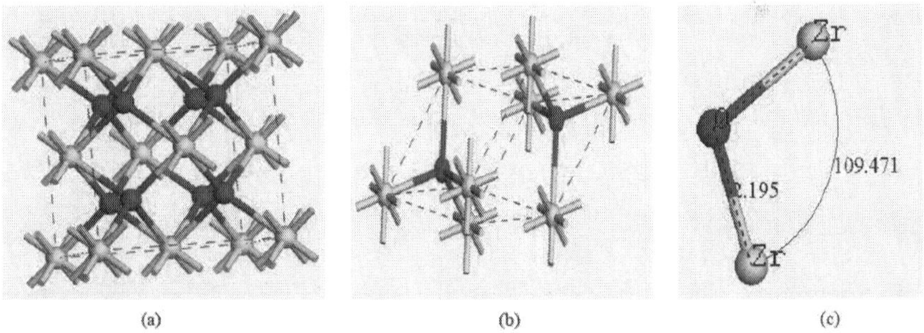

Figure 1. Crystal structure of $cZrO_2$ in various forms: (a) Unit Cell; (b) Primitive Cell; (c) Zr-O Bond (O_2 and Zr atoms are represented by red and light blue balls respectively).

Table 1. Lattice parameters of optimized $cZrO_2$.

Parameters	Simulations		
	Present Study	Other(s)	Experimental
a = b = c (Å)	3.55381	3.58031 [4]	3.6012 [17]
Bond Length (Zr-O), Å	2.1763	2.19537 [2]	2.205 [5]
Volume (Å^3)	31.8808	32.5850 [3]	32.723 [17]
Density (g/cm^3)	6.41791	6.081 [18]	6.151 [18]

3.2. Mechanical Properties

The mechanical properties of $cZrO_2$ are based on the elastic constants of the crystal which indicates response to external forces. There are six components of stress and a corresponding six components of strain for the general 3-D case. Thus, Hooke's law may be expressed as:

$$\sigma_i = C_{ij}\varepsilon_j \tag{1}$$

$$\varepsilon_i = S_{ij}\sigma_j \tag{2}$$

where C = stiffness or elastic constant, S = compliance, σ = stress and ε = strain. In matrix format, the stressstrain relation showing the 36 (6 × 6) independent components of stiffness can be represented as [3] :

$$\begin{pmatrix} \sigma_1 \\ \sigma_2 \\ \sigma_3 \\ \sigma_4 \\ \sigma_5 \\ \sigma_6 \end{pmatrix} = \begin{pmatrix} C_{11} & C_{12} & C_{13} & C_{14} & C_{15} & C_{16} \\ C_{21} & C_{22} & C_{23} & C_{24} & C_{25} & C_{26} \\ C_{31} & C_{32} & C_{33} & C_{34} & C_{35} & C_{36} \\ C_{41} & C_{42} & C_{43} & C_{44} & C_{45} & C_{46} \\ C_{51} & C_{52} & C_{53} & C_{54} & C_{55} & C_{56} \\ C_{61} & C_{62} & C_{63} & C_{64} & C_{65} & C_{66} \end{pmatrix} \begin{pmatrix} \varepsilon_1 \\ \varepsilon_2 \\ \varepsilon_3 \\ \varepsilon_4 \\ \varepsilon_5 \\ \varepsilon_6 \end{pmatrix} \tag{3}$$

The number of independent elastic constants in $cZrO_2$is three (3), i.e. C_{11}, C_{12} and C_{44} which are computed as 596.33, 137.04 and 74.34 GPa respectively. Also the bulk modulus (B) was calculated as 290.134 GPa using the Voigt model, which is comparable to experimental value [17] .

With reference to Born-Huang's lattice dynamic theory [19] , the mechanical stability of a cubic system is based on the conditions expressed in Equations (4) and (5).

$$C_{11} > 0, C_{44} > 0, C_{11} - C_{12} > 0, C_{11} + 2C_{12} > 0 \quad (4)$$

$$C_{12} < B < C_{11} \quad (5)$$

Based on conditions outlined in Equations (4) and (5), the structure is mechanically stable due to lattice parameters and orientation of the crystal. The stability will make polycrystalline $cZrO_2$ less vulnerable to generating micro cracks [19] .

For $cZrO_2$, computed Young's modulus (E) is 545.12 GPa, shear modulus (G) based on Voigh model is 136.464 GPa and the Lame modulus based on Reus model is 222.195 GPa. The G/B ratio indicates the ductility or brittleness of a material [19] and was found out to be 0.469, which indicates that $cZrO_2$ ionic and brittle. This is becausefor covalent and ionic materials, the typical relations between bulk modulus and shear modulus are as $G \approx 1.1B$ and $G \approx 0.6B$, respectively[20] . The Poisson's ratio (υ) of $cZrO_2$ vary from 0.14 to 0.189, which indicates less metallic and ionic character in the Zr-O bond.

In crystals, hardness quantifies the resistance to deformation and may be predicted using microscopic models [17] . The hardness (Hv) of $cZrO_2$ is determined using a semi-empirical equation defined as follows [21] [22] :

$$H_v = 0.92K^{1.137}G^{0.708}, K = G/B \quad (6)$$

The obtained hardness (Vickers number) for $cZrO_2$ is 12.663, which is consistent with the chemical bonding and elastic modulus as analyzed above and in similar other studies [19] -[22] .

3.3. Thermodynamic Properties

The thermodynamic properties of a solid have direct relevance to its phonon characteristics as it indicates the quantum of elastic strain energy [19] . Hence, the phonon dispersion curve of a material elucidates the thermal properties based on the concept of lattice vibrations and interpretation of lattice dynamics [15] . Using the method of linear response with 0.05 1/Å q-vector grid spacing, the phonon dispersion curves (PDC) for $cZrO_2$ was generated as shown in Figure 2. The lower section of G-X direction indicates higher longitudinal energy while most of the branches in the X-K direction converge as non-degenerate, thus leading to the expected nine branches for a typical cubic structure [19] .

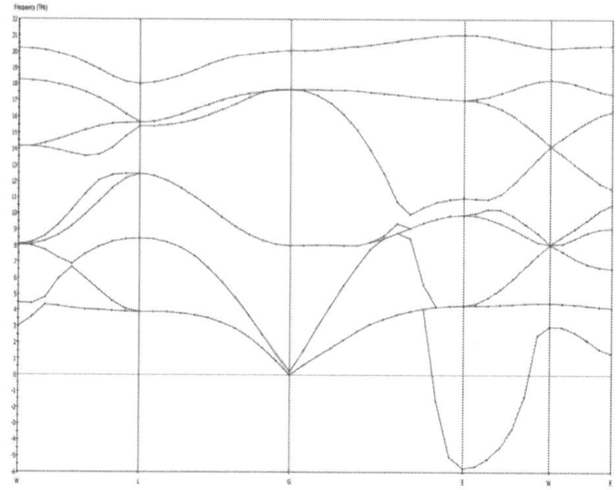

Figure 2. Phonon dispersion curves of $cZrO_2$.

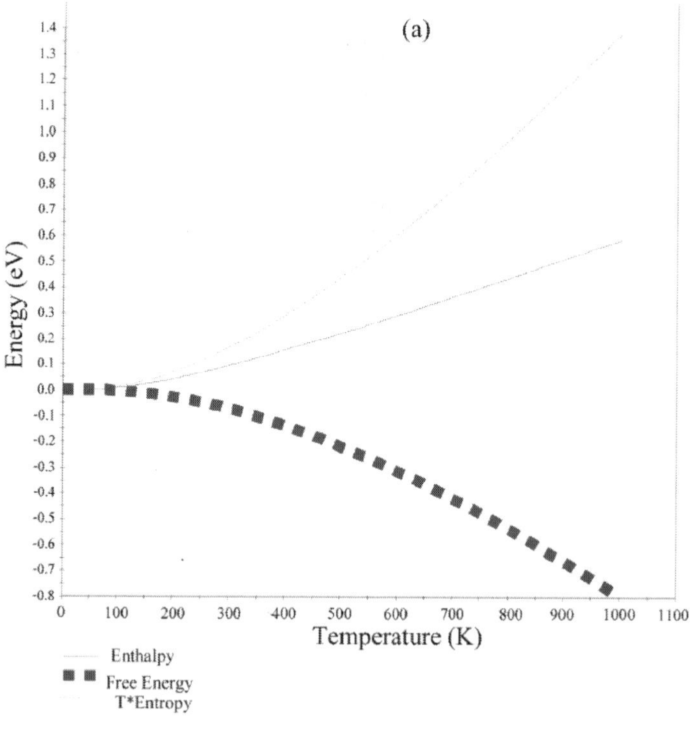

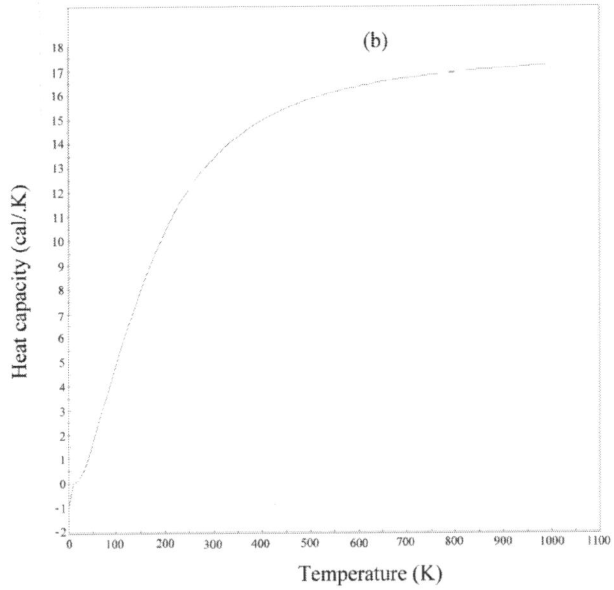

Figure 3. Enthalpy, free energy and entropy (a) and heat capacity of $cZrO_2$.

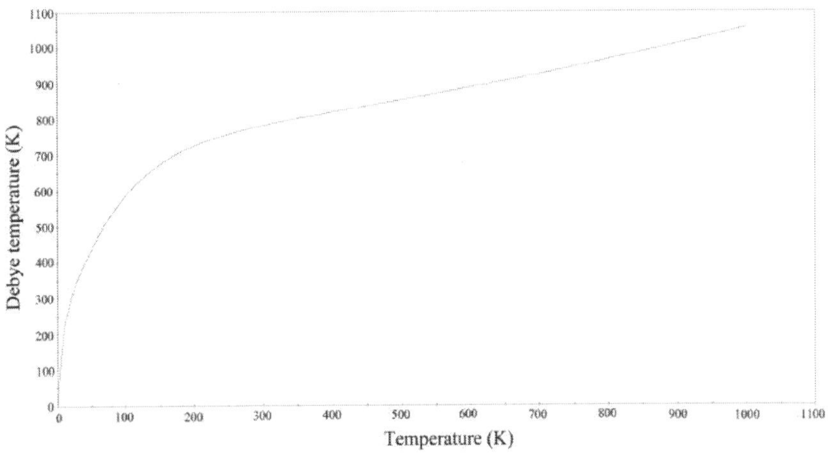

Figure 4. Debye temperature graph of $cZrO_2$.

With the aid of data from the phonon dispersion curves (PDC), quasi-harmonic approximation was used to evaluate some temperature dependent properties of $cZrO_2$ such as entropy, enthalpy, free energy and Debye temperature as illustrated in Figure 3 and Figure 4. From Figure 3, the zero-point energy was computed to be 0.20385 eV and the heat capacity per unit cell had a maximum value of 16.8957 cal/K (0.07069 kJ/K) at 992.25 K. Debye temperature gives an approximation for the low temperature heat capacity of insulating crystalline solids. The maximum Debye temperature for $cZrO_2$ was found to be 946 K which is similar to 963 K for Y_2O_3 stabilized $cZrO_2$ determined experimentally using neutron powder diffraction [23] . All the computed thermodynamics properties of $cZrO_2$ indicates it has less thermal conductivity despite having high ionic conductivity thus confirming its major characteristic as a refractory material.

4. CONCLUSION

Using the first principles calculations, the properties of $cZrO_2$ were investigated. The calculated lattice parameters showed conformity with available experimental data. The computed properties based on elastic constants

indicate that $cZrO_2$ has satisfied the mechanical stability requirements; however, the young's modulus and hardness are high when compared to similar material such as cerium oxide. As expected, the thermodynamic properties obtained confirmed the ceramic characteristics of $cZrO_2$. The data obtained may further be used to predict properties in relation to defects and/or dopants and setting parameters for processing nanocomposites related to $cZrO_2$.

ACKNOWLEDGEMENTS

The authors are grateful for the supports provided by Universiti Teknologi Petronas and Malaysian Ministry of Higher Education (MOHE) through the Long Term Research Grant Scheme (LRGS) for One Baja Research Programme (Project 6).

REFERENCES

1. Xia, X., Oldman, R. and Catlow, R. (2009) Computational Modeling Study of Bulk and Surface of Yttria-Stabilized Cubic Zirconia. Chemistry of Materials, 21, 3576-3585.

2. Bandura, A.V. and Evarestov, R.A. (2012) Ab Initio Structure Modeling of ZrO2 Nanosheets and Single-Wall Nanotubes. Computational Materials Science, 65, 395-405.

3. Wang, C. (2009) Multiscale Modeling and Simulation of Nanocrystalline Zirconium Oxide. Ph.D. Thesis, University of Nebraska at Lincoln.

4. Muhammad, I.D. and Awang, M. (2013) Modelling the Interatomic Potential of Cubic Zirconia. Applied Mechanics and Materials, 446-447, 151-157.

5. Suciu, C., Gagea, L., Hoffmann, A.C. and Mocean, M. (2006) Sol-Gel Production of Zirconia Nanoparticles with a New Organic Precursor. Chemical Engineering Science, 61, 7831-7835.

6. Takenaka, S., Uwai, S., Ida, S., Matsune, H. and Kishida, M. (2013) Bottom-Up Synthesis of Titania and Zirconia Nanosheets and Their Composites with Graphene. Chemistry Letters, 42, 1188-1190.

7. Kulkova, S., Bakulin, A., Hocker, S. and Schmauder, S. (2012) Ab-Initio Study of Metal-Zirconia Interfaces. Materials Science and Engineering, 38, 012004.

8. Lamperti, A., Cianci, E., Ciprian, R., Sangalli, D. and Debernardi, A. (2013) Stabilization of Tetragonal/Cubic Phase in Fe Doped Zirconia Grown by Atomic Layer Deposition. Thin Solid Films, 533, 83-87.

9. Welberry, T.R., Withers, R.L., Thompson, J.G. and Butler, B.D. (1992) Diffuse Scattering in Yttria-Stabilized Cubic Zirconia. Journal of Solid State Chemistry, 100, 71-89.

10. Hou, Z.F. (2008) Ab Initio Calculations of Elastic Modulus and Electronic Structures of Cubic CaZrO3. Physical B: Condensed Matter, 403, 2624-2628.

11. Zhang, P., Lu, Y., He, C. and Zhang, P. (2011) First-Principles Study of the Incorporation and Diffusion of Helium in Cubic Zirconia. Journal of Nuclear Materials, 418, 143-151.

12. Zhao, X., Shang, S., Liu, Z. and Shen, J. (2011) Elastic Properties of Cubic, Tetragonal and Monoclinic ZrO2 from First-Principle's Calculations. Journal of Nuclear Materials, 415, 13-17.

13. Miller, S.P., Dunlap, B.I. and Fleischer, A.S. (2012) Cation Coordination and Interstitial Oxygen Occupancy in Co-Doped Zirconia from First Principles. Solid State Ionics, 227, 66-72.

14. Accelrys Software Inc., San Diego (2012) Materials Studio.

15. Clark, S.J., Segall, M.D., Pickard, C.J., Hasnip, P.J., Probert, M.J., Refson, K. and Payne, M.C. (2005) First Principles Methods Using CASTEP. Zeitschrift Fuer Kristallographie, 220, 567-570.

16. Perdew, J.P., Burke, K. and Ernzerhof, M. (1996) Generalized Gradient Approximation Made Simple. Physical Review Letters, 77, 3865-3868.

17. Soo, Y.L., Chen, P.J., Huang, S.H., Shiu, T.J., Tsai, T.Y., Chow, Y.H., et al. (2008) Local Structures Surrounding Zr in Nanostructurally Stabilized Cubic Zirconia: Structural Origin of Phase Stability. Faculty Publications—Chemistry Department, Paper 18.

18. Chang, Y., Wang, H., Zhu, Q., Luo, P. and Dong, S. (2013) Theoretical Calculation and Analysis of ZrO2 Spherical Nanometer Powders. Journal of Advanced Ceramics, 2, 21-25.

19. Goldsby, J.C. (2013) Basic Elastic Properties Predictions of Cubic Cerium Oxide Using First-Principles Methods. Journal of Ceramics, 2013, Article ID: 323018

20. Yang, Z.-J., Guo, Y.-D., Linghu, R.-F. and Yang, X.-D. (2012) First-Principles Calculation of the Lattice, Compressibility, Elastic Anisotropy and Thermodynamic Stability of V2GeC, China Physics B, 21, 036301

21. Tian, Y., Xu, B. and Zhao, Z. (2012) Microscopic Theory of Hardness and Design of Novel Superhard Crystals. International Journal of Refractory Metals and Hard Materials, 33, 93-106.

22. Chong, X., Jiang, Y., Zhou, R. and Feng, J. (2014) First Principles Study the Stability, Mechanical and Electronic Properties of Manganese Carbides. Computational Materials Science, 87, 19-25.

23. Kisi, E. and Yuxiang, M. (2003) Debye Temperature, Anharmonic Thermal Motion and Oxygen Non-Stoichiometry in Yttria Stabilized Cubic Zirconia. Journal of Physics: Condensed Matter, 10, 3823-3832.

Index